International Tourism Futures

The Drivers and Impacts of Change

International Tourism Futures

The Drivers and Impacts of Change

Clare Lade, Paul Strickland, Elspeth Frew, Paul Willard, Sandra Cherro Osorio, Swati Nagpal and Peter Vitartas

(G) Goodfellow Publishers Ltd

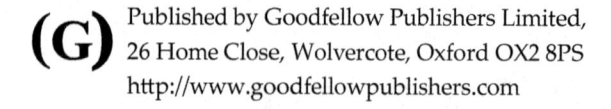

(G) Published by Goodfellow Publishers Limited,
26 Home Close, Wolvercote, Oxford OX2 8PS
http://www.goodfellowpublishers.com

British Library Cataloguing in Publication Data: a catalogue record for this title is available from the British Library.

Library of Congress Catalog Card Number: on file.

ISBN: 978-1-911635-23-9

DOI: 10.23912/9781911635222-4383

Copyright © Clare Lade, Paul Strickland, Elspeth Frew, Paul Willard, Sandra Cherro Osorio, Swati Nagpal and Peter Vitartas, 2020

All rights reserved. The text of this publication, or any part thereof, may not be reproduced or transmitted in any form or by any means, electronic or mechanical, including photocopying, recording, storage in an information retrieval system, or otherwise, without prior permission of the publisher or under licence from the Copyright Licensing Agency Limited. Further details of such licences (for reprographic reproduction) may be obtained from the Copyright Licensing Agency Limited, of Saffron House, 6–10 Kirby Street, London EC1N 8TS.

All trademarks used herein are the property of their repective owners, The use of trademarks or brand names in this text does not imply any affiliation with or endorsement of this book by such owners.

Design and typesetting by P.K. McBride, www.macbride.org.uk

Cover design by Cylinder

Printed by Marston Book Services, www.marston.co.uk

Contents

1	Introduction	1
2	Drivers of Change	7
3	Tourists of the Future	21
4	Hospitality of the Future	39
5	The Future of Visitor Attractions	55
6	Events of the Future	73
7	The Future of Tourism, Hospitality and Events Teaching and Training	87
8	The Future of Film Tourism	103
9	The Future of Health and Wellness Tourism	117
10	Sustainable Development and Responsible Tourism	135
11	Future Proofing a Crisis	149
12	Solving Future Problems in the Tourism, Hospitality and Events Sectors	171
13	The Demise of Tourism?	187
14	Building Future Scenarios	205
15	Summary	223
	Index	227

About the authors

Dr Elspeth Frew is an Associate Professor in Tourism, Hospitality and Event Management in the La Trobe Business School. Elspeth's research interest is in cultural tourism, with a particular focus on dark tourism, industrial tourism and festival, event and attraction management. She has published work in these areas as journal articles, book chapters and edited books – most recently on dark tourism and the relationship between tourism and national identities.

Dr Clare Lade is a lecturer in the Bachelor of Hospitality Management at Melbourne Polytechnic. She has held previous lecturing positions in tourism, hospitality and events at both La Trobe University and Monash University, Victoria, Australia. Her research interests include regional tourism development, dark tourism, gastronomic tourism, festivals, and events.

Dr Swati Nagpal is a lecturer at La Trobe University. Her research and teaching focuses on corporate social responsibility and sustainability and has published and been involved in a number of sustainability and CSR-related research projects. These include research on sustainability in higher education, CSR transformation in Australian ASX 200 companies, sustainable procurement in Australian and UK universities, understanding social risk, and a study of Australian corporate responses to climate change.

Dr. Sandra Cherro Osorio is Head of Program in the Bachelor of Hospitality Management at Melbourne Polytechnic. Sandra has extensive industry experience in tourism and hospitality from different countries including Peru, the United States, the United Kingdom and Australia. Her research interests include gastronomy, tourism, community development, and higher education.

Paul Strickland is a Lecturer at La Trobe University specialising in Hospitality Management subjects. Paul has a vast background of job titles in industry including hotel and restaurant management roles in many countries. His research interests include food, wine and space tourism and specialised in ethnic restaurants for post graduate studies. He also teaches in a Hospitality Management program in Bhutan.

Dr Peter Vitartas is a marketing academic who has taught at a number of higher education institutions both in Australia and overseas. His research and teaching focuses on social marketing and public policy, community and economic development and learning analytics in higher education. He has published extensively and presented at national and international conferences on topics covering marketing management and public policy, customer satisfaction, media and time use, services and direct marketing, tourism and business education.

Dr Paul Willard lectures in tourism, hospitality and events at La Trobe University at the Bendigo campus in central Victoria. Paul has an extensive background in the hospitality and events industry having worked under various job titles. His research interests include heritage tourism, cycling tourism, experiential tourism and event management.

1 Introduction

Introduction

> Education is our passport to the future, for tomorrow belongs to the people who prepare for it today.
>
> Malcolm X

This co-authored book was researched and written during a time that few had foreseen, let alone prepared for. The impacts of Covid-19 are being felt across the world's societies, economies and natural environment. Some industries have been more impacted than others, including the international tourism industry. The United Nations World Tourism Organisation (UNWTO) predicts that due to the travel related impacts of Covid-19 international tourism could decline by between 60-80% in 2020, with US$80 billion already lost in exports from the industry for the first quarter of 2020 (UNWTO, 2020a).

In these unprecedented times, it becomes more important than ever to consider what the future might hold for the industry. By examining current and future capabilities of the industry, this research book explores the opportunities available to shape the future through rebuilding, disrupting and developing greater resilience in the tourism industry. The common theme throughout the chapters is change – no matter how change emerges, the authors of this book recognise that the industry is always going to face times of turbulence, whether it be climate change, political or financial disruptions or pandemics, those in the industry need to have resilience, understand the forces of change and be prepared to adapt. This chapter sets out the core principles associated with anticipating the future of the international travel, hospitality and events sectors. It starts with a broad overview of the global tourism industry, followed by the definitions and scope of the sectors that will be covered in the book. A discussion on tourism futures as an area of research is presented and finally, the sections and individual chapters are introduced.

The global tourism industry today

Notwithstanding the current threat to the industry from Covid-19, tourism has seen significant growth in the last few decades based largely on the rising disposable income and living standards in a significant part of the world's population. To take advantage of this growth in leisure and tourism expenditure, it is important that businesses, governments and other stakeholders in the entire tourism ecosystem understand the trends that are likely to shape the future of the industry (Buhalis *et al.*, 2006).

The definition of the tourism industry for the purposes of this book has been adapted from the UNWTO as being:

> The cluster of production units in different sectors that provide
> consumption goods and services demanded by tourists. Such
> clusters are called sectors because tourist acquisition represents such
> a significant share of their supply that, in the absence of tourists,
> their production of these would cease to exist in meaningful quantity
> (UNWTO, 2008).

The three sectors referred to in the above definition that form a key focus of this book are the travel, hospitality and events sectors. Therefore, when referring to the 'tourism industry', all three sectors are encompassed in the discussion although we also recognise that the industry is sometimes referred to as the 'THE' sector.

It is important to note that the drivers of transformation in the industry continue to arise from both the demand and supply side. On the demand side, we have seen rising disposable income, favourable exchange rates, technological change, government regulations and changing consumer preferences for leisure activity and spending (Page and Connell, 2020). In response, governments and local tourism operators have moved to step in and meet this growing demand. Supply is rising to meet this increased demand in relation to transport, accommodation, visitor-attractions, associated services and government incentives to attract tourists (Dwyer *et al.*, 2020).

Governments at all levels – national, regional and local, have grown to appreciate the value of tourism income for their economies and the importance it has made to the employment and training of people. World-wide tourism arrivals in 2019 included 1.5 billion travellers generating US$1.5 billion (UNWTO, 2020e). Its impact cannot be understated with almost all countries in the world depending to some extent on tourists to support their economy. This highlights the importance of the industry and recognises that it is in the interest of all countries that tourism continues to exist globally in some capacity. Aside from the economic and employment benefits there are many other important contributions that tourism makes. Whether it be raising awareness of the many cultural differences around the world, highlighting the importance of national treasures or simply bringing people together so that they might experience conversations, stories or understand the many ways that people live, tourism makes for a much richer and harmonious globe.

Why tourism futures?

With the growth and size of the global tourism industry, it becomes important to understand and effectively frame the future of the industry for business leaders and policy-makers (Veal *et al.*, 2015; Yeoman and Postma, 2014). Tourism futures is a relative new area that is evolving, and involves analysing trends, patterns, historical underpinnings and change to compose a range of possibilities for the industry in the medium to long term future (Buhalis *et al.*, 2006).

Yeoman and Beeton (2014: 676) stated that 'Given the multidisciplinary nature of the field of tourism futures, it is not founded on one theoretical position.' Thus, there are three features of futures studies that set the tone of this book. First, we aim to take a systems' view of the industry, which includes a consideration of the range of stakeholders including, *inter alia*, consumers, producers, governments, NGOs and local communities. This allows for a holistic understanding of the scope of industry, and important inter-relationships between stakeholders that might shape future scenarios. Second, we examine potential and probable future trends based on an analysis of the socio-cultural technological, economic, environmental, political and international dimensions. Third, we take a medium to long-term view of the future potential and opportunities available to the travel, hospitality and events sectors that extends beyond the one to three years strategic planning cycle.

Included in our discussions is consideration of a vast number of theories and concepts which inform the study of tourism futures. These topics, as they are explored in the chapters are summarised in Table 1.1 and highlight the breadth and depth of discussion in the chapters. While reoccurring themes include globalisation, sustainability, artificial intelligence, problem solving, together the theme of change and changes for the future weave their way through the chapters.

Table 1.1: Chapter summary

Chapter	Chapter Title	Concepts Discussed
1	Introduction	Sets the theme of change
2	Drivers of Change	Leisure, Globalisation, Climate change, Segmentation, Collaborative consumption, Consumer activism
3	Tourists of the Future	Smart tourism, Virtual tourism, Super sabbaticals, Solo travellers
4	Hospitality of the Future	Space hotels, Underwater accommodation, Cruise ships, Robots in hospitality, Artificial intelligence
5	Visitor Attractions	Theme parks, Visitor attractions, Over-tourism
6	Events	Event formation, Event sustainability, MICE, Virtual events, Safety and Security
7	Teaching and Training	Contemporary teaching, MOOCs, Open badges
8	Film Tourism	Film tourism and destinations, Challenges of film tourism, Future film tourists
9	Health and Wellbeing	Definitions of Health, Medical, Wellness and Spa tourism, Health transformations, Wellness providers, Megatrends
10	Sustainable Development	Sustainability and tourism, Sustainable development goals, three pillars of sustainability, Responsible tourism, Long-term systems approaches
11	Futureproofing a Crisis	Crisis definitions, Impacts of crises, Impacts of terrorism, Impacts of natural disasters, Impacts of disease outbreaks, Crisis management models, Destination crisis management
12	Solving Future Problems	Traditional approaches to problem solving, Routine problem solving, Non-linear solutions, Wicked problems, Solving complex problems, Systems thinking, Systems design, Causal mapping, Stocks and flows, Scenario planning, Frameworks for the future
13	The Demise of Tourism	Plausible futures, Natural resources, Minerals, Fuels, Agricultural resources, Morality and ethical concerns, Impact of war, Global population growth, Food security
14	Building Future Scenarios	15 scenarios for the future
15	Summary	

Organisation of the book

The book is divided into four sections, namely:
1 Futures;
2 Sectors;
3 Themes;
4 Foresight.

Futures

In Chapter 2, we explore the drivers that are likely to influence tourism in the future, through an analysis of the external and internal forces of change in the industry. Concepts such as smart tourism, virtual tourism, underwater hotels, artificial intelligence are introduced to showcase examples of what is currently being implemented in some countries and/or sectors. Other concepts such as the potential of space tourism and impacts of climate change are also explored to indicate how the industry may change depending on demand. This is followed by a detailed look into the rise of the middle class in the decades leading up to 2050 in Chapter 3.

Sectors

Hospitality trends and future scenarios are discussed in Chapter 4, highlighting innovative products and services that are becoming available. In Chapter 5, we explore the future of visitor attractions by using the examples of theme parks and world-renowned tourist attractions to illustrate the possible developments in this area, and we consider the impact of social media and over-tourism on these attractions. Events of the future are covered in Chapter 6, where we delve into the various contemporary issues which are likely to make an impact on events in the future which include event sustainability, event inclusivity and event technology.

Themes

In order to meet the future education and training needs of the industry, Chapter 7 explores the development of education in tourism, hospitality and events, and the contemporary factors influencing learning and teaching. In Chapter 8, the focus is turned to the future of film tourism, which continues to grow in popularity. Similarly, wellness tourism is currently one of the fastest growing tourism niche markets, which is covered in Chapter 9. In Chapter 10, we consider the tourism industry's impacts on socio-cultural, environmental and economic dimensions of sustainable development into the future.

Key definitions relating to crisis management are identified and a disaster management model introduced in Chapter 11. Discussion in this chapter is largely within the context of terrorism and natural disasters with a number of examples used to demonstrate its application.

Foresight

In the remaining chapters, our attention is turned to the development of foresight among tourism practitioners, students and policy-makers. There is a focus on future approaches to problem-solving in Chapter 12, presenting potential scenarios that spell the demise of tourism as we know it in Chapter 13, and building of future scenarios in the travel, hospitality and events sectors in Chapter 14. Finally, a summary of the book is presented in Chapter 15.

Each chapter concludes with a case study to demonstrate in a practical manner the core elements discussed within the chapter. Offering case studies which highlight practical implications makes for an ideal resource for teaching in tourism studies and to evoke reflection, innovation and change as a means to enhance the tourism industry as a large and diverse sector.

References

Buhalis, D., Costa, C. and Ford, F. (2006) *Tourism business frontiers*, Oxford: Routledge.

Dwyer, L., Forsyth, P. and Dwyer, W. (2020) *Tourism Economics and Policy*, Bristol: Channel View Publications.

Page, S. J. and Connell, J. (2020) *Tourism: A modern synthesis*, Oxford: Routledge.

UNWTO. (2008) *Understanding Tourism: Basic glossary*, n.d., viewed 20/05/2020, http://media.unwto.org/en/content/understanding-tourism-basic-glossary

UNWTO (2020a) *International tourist numbers could fall 60-80% in 2020*, 7 May, viewed 20/05/2020 https://www.unwto.org/news/covid-19-international-tourist-numbers-could-fall-60-80-in-2020

UNWTO. (2020e) *International tourism growth continues to outpace the global economy*, January 2020, viewed 15/07/2020, https://www.unwto.org/international-tourism-growth-continues-to-outpace-the-economy

Veal, A.J., Darcy, S. and Lynch, R. (2015) *Australian Leisure*, 4th edn, Frenchs Forest: Pearson Higher Education.

Yeoman, I.S. and Beeton, S. (2014) 'The state of tourism futures research: An Asian Pacific ontological perspective.' *Journal of Travel Research*, **53** (6), 675-679.

Yeoman, I. and Postma, A. (2014) Developing an ontological framework for tourism futures. *Tourism Recreation Research*, **39** (3), 299-304.

2 Drivers of Change

Introduction

At first glance, it would appear that tourism is ubiquitous; it is a global activity that is experienced in all countries, with every country having an equal stake. However, this is misleading as it has been demonstrated that international tourism is 'dominated by relatively few countries', with tourism being described as an activity 'open to an elite only'. As such, tourism has traditionally been engaged in by those who are from prosperous countries, who have higher incomes and stable and secure societies (Todd, 2001: 12). International tourism is dominated by the 'wealthier, industrialised world' (Sharpley, 2018: 50), with the major tourism flows occurring between the more developed countries or from developed countries to developing countries. In recent years some new destinations such as Brazil, Russia, India, China and South Africa have emerged (Weaver and Lawton, 2014), challenging the dominance of the traditional generating and receiving countries. However, the majority of travel is still undertaken by tourists from traditional tourism generating regions of Europe, North America and parts of South East Asia (Sharpley, 2018: 50).

Despite this concentration of tourist origins, many countries around the world are interested in receiving tourists for the economic benefits including local employment and foreign currency. Todd (2001: 14) suggests that if every country in the world had 'peace with its neighbours and a rational approach to development', they could see tourism play its part in developing their economies. From a demand side perspective, 'rising real incomes, expanding discretionary spending, increasing leisure time, faster and cheaper transport and the spread of global awareness through the printed and broadcast media and… through the internet' fuel tourism growth (Todd, 2001: 15). But the question remains, what will happen to tourism in the future?

Predicting the future

Tourism and leisure forecasting has been described as being more similar to economic forecasting than to weather forecasting because it involves a human element. As such, this makes attempting to foresee the future in tourism 'highly speculative' (Veal *et al.*, 2015: 512). However, a good way to start any examination of future travel is to reflect on our understanding of the past and the present as this can help us to speculate on the future based on a 'dispassionate viewing of recent trends' (Veal *et al.*, 2015: 512). Sharpley (2018: 52) suggests that the future is both easy and difficult to predict because it is impossible to know if a prediction is accurate until 'the future becomes the present'. On the other hand, it is impossible to know which factors might influence the future of tourism as they can range from technological innovation to political or economic events.

For tourism to occur an individual has to have the time, money and means to travel (Weaver and Lawton, 2014). As society develops, a larger proportion of the world's population has access to leisure time and a higher disposable income. There is also increased awareness of tourist attractions and destinations which people seek to visit (Morrison, 2019). Couple this with relatively cheap airfares and these are the perfect conditions for tourism to grow, as illustrated in the last 50 years (Page, 2019). It can also be expected there will be corresponding increases in hospitality as outlined in Chapter 4. Similarly, the frequency and variety of events staged across the world has increased and contributed to the growth of tourism (Getz, 2008). A networked global economy, driven by 'rapid and largely unrestricted flows of information, ideas, cultural values, capital, goods and services, and people (for example globalisation)' (Dwyer, 2012: 531) has ensured that citizens will seek to visit and experience attractions and destinations.

Leisure time

There are many, and often contradictory, arguments used to predict the future of tourism based on the relationship between work and leisure. This is because, in this post-industrial world where rigid patterns of work have been challenged, leisure time and availability to engage in tourism has changed. Such a blurring of work and leisure time reflects the diversified location of the workplace, the expansion of part-time and casual work and the flexibility in working hours changing career opportunities. As a result, people no longer necessarily regard the weekend as a time of rest and relaxation: weekend shopping is now the norm, and people fit their leisure and tourism activities around the seven-day week, 24-hour economy (Veal *et al.*, 2001). These shorter, more frequent and more intensive periods of leisure spread evenly throughout the year allows stressed workers to get away from their work to rest and recover.

It is difficult to predict whether individuals in developed countries will have more or less leisure time in the future because in some countries, the retirement age will rise and for others it will fall (Veal *et al.*, 2001). For wealthier retirees, leisure time is abundant and there is an increasing proportion of people aged over 60 years who are 'retiring earlier, fitter and with wide leisure interests' (Clark, 2001: 76). This ageing population has an interest in travel and tourism leading to an opportunity for tourism operators to cater for this growing market sector. Consequently there has been growing demand for 'health tourism and wellness activities' (Sharpley, 2018: 54) as discussed further in Chapter 9 of this volume.

Competition in the world economy and increased life expectancy means that some governments are reducing pensions and raising the retirement age. The affected population require increased pension contributions and more savings for their retirement. This leads to employees becoming worried about employment security and their income which could force them to remain in the workforce longer 'rather than thinking about long holidays' (Clarke 2001: 77). An increase in the retirement age may reflect the so-called 'dependency ratio', which is a measure of the number of economically inactive people (such as children, retired, unemployed), relative to those in the work-force. This can place unsustainable demands on public finances resulting in lower pensions, higher taxes, and people working longer. As a consequence, the younger and older generations in Western countries by the mid-21st century may have 'much lower disposable income and, hence, spend less on tourism' (Sharpley, 2018: 54). In effect, the leisure and tourism market are becoming increasingly segmented into those people who have money but are time-poor, and those who have less money but are time-rich. The time-poor tourists are likely to seek products that are tailored for a tight schedule to make the most of their limited leisure time. They may buy upmarket packages which allow them to be pampered, are exotic and fashionable but are for a shorter period of time. In contrast, time-rich tourists who are free of time constraints and with less disposable income may be 'open to enjoying several holidays each year', or extended trips which are inexpensive and more affordable (Clark, 2001: 79). The development of high-speed trains and further growth in low-cost airline operations will allow shorter breaks and excursions, as well as city tourism to become increasingly popular (Sharpley, 2018).

An alternative view on leisure is presented in West's book on the future of work (2018: 84). He suggests that one of the benefits of the way that work is evolving is that people will have more leisure time than in the past, since some people will 'not be needed in the new digital economy and so they will find other ways to construct meaning in their lives outside the workplace'. In addition, those who work may find themselves with time on their hands to pursue other kinds of non-work activities including art, culture, music, sports, theatre and 'a range of interests' (West, 2018: 85). This supports the philosophical concept of 'work in order to play', whereby we work to obtain the necessary funds to undertake worthwhile leisure pursuits. This is a post-industrial idea which was derived from ancient Greek philosophy where leisure was viewed as being important in its own right (Weaver and Lawton, 2014).

Whether or not leisure will increase or decrease, the existence of leisure and associated tourism activities will continue to play an important role in society. People will seek ways to relax and recharge before returning to work, with work being paid or voluntary, full-time, part-time, casual, sessional, based in traditional offices, home-based or a hybrid of locations.

Human nature and future travel

Humans are 'animated by a desire for new experience' (McCannell, 2013: xxii). From a tourist perspective this implies that tourists may be interested in visiting destinations where they can experience the architecture, the nightlife, food, culture and the people. De Botton (2002) suggests that travelling provides an understanding of what life might be about 'outside the constraints of work and the struggle for survival' (p.9), and proposes that travel can be associated with the exotic in that 'the charm of a foreign place arises from the simple idea of novelty and change' (p.78). According to De Botton there may be a more profound pleasure to be found in travel whereby 'we may value foreign elements not only because they are new, but because they seem to accord more faithfully with our identity and commitments than anything our homeland could provide' (p.78). He suggests that what we find exotic abroad may be what 'we hunger for in vain at home' (p.78). He goes on to argue that nature is something which encourages us to travel, and that vast spaces of nature perhaps provide us with the 'finest, the most respectful reminder of all that exceeds us' (p.178). De Botton points out that 'we tend to seek out corners of the world only once they have been painted and written about by artists' (p.214).

Based on De Botton's thinking about human nature, we could predict that tourists of the future are likely to travel to experience novelty and change away from their home environment with particular desires to experience nature and sites of cultural experience. As a result, in the future people will be interested in visiting countries with strong natural or cultural identities demonstrated through unusual geography, habitat or historical significance and featuring exotic food, drink and customs. This view reinforces that pristine natural areas of the world should be protected if, for nothing else, tourists to experience. In the future there may be a 'rejection of the traditional sun-sea-sand holidays in favour of healthier, more active holidays' (Sharpley, 2018: 54), where tourists travel to explore the novel both in nature and culture. It can be expected there will be greater demand for educational or cultural experiences, reflecting a desire to learn about other cultures, and products that are more tailored to individual lifestyles which moves away from the mass tourism products introduced in the 1970s and 1980s (Sharpley, 2018). This later topic is discussed in more detail in Chapter 3.

Globalisation and climate change

Globalisation has been strongly associated with tourism as it has encouraged international trade, travel and informed people about tourist destinations (Dwyer, 2015). Aviation developments have supported globalisation and international tourism with transport playing a crucial role in moving large numbers of people effectively from their home environment to the tourist destination. Alexandre De Juniac, the CEO of the International Air Transport Association which represents 82% of global air traffic, stated that he is against a movement to stop or heavily reduce flying because that would have grave consequences for people, jobs and economies around the world. According to De Juniac, that would be a 'step backward to an isolated society that is smaller, poorer and constrained'. He suggested that flying has 'made the world a better, freer place and we are committed to sustainably making it better and even freer' (De Juniac, 2019). He also noted that since 1988 the number of unique city pairs (for example, air flights between pairs of popular cities) around the world has doubled to 22,375 with flying becoming more affordable and accessible to more people with an average return airfare being cheaper than those in 1998, after adjusting for inflation (De Juniac, 2019). Indeed the reduction in tourists travelling internationally due to Covid-19 has had a serious impact on the airline industry with some commentators stating that the coronavirus pandemic has 'reset the clock on a decades-long aviation boom that's been one of the great cultural and economic phenomena of the postwar world' (Whitley, 2020). The severe reduction in demand has led to some airlines (such as Virgin Australia) closing down and others severely reducing their flights. However, once the coronavirus pandemic has eased and travel bans are lifted then individuals may desire to visit family and friends once again, with the expectation that all air travel and airports operate with good hygiene (Whitley, 2020).

Due to the coronavirus pandemic, commentators believe there may be a re-thinking of essential travel for business, with video conferencing replacing the need to fly by many business travellers (Whitley, 2020). In addition, there is a growing awareness of the impact of air travel on climate change and a movement has emerged to reduce air travel in order to minimise the damage on the environment (Cocolas et al., 2020). The evidence that climate change is already impacting destinations includes extreme weather events and decreased snow cover in ski resorts, with tourism generating greenhouse gases via air travel, car and bus emissions (Sharpley, 2018). For example, the 2019/2020 climate change related bushfires in Australia have led to many tourists deciding not to visit these bushfire impacted areas, despite the fires being extinguished. In 2019 there was a shift in public opinion in relation to sustainability in general and climate change in particular. This change in public opinion was encouraged and supported by climate change activist groups such as Extinction Rebellion, which have organised global climate strikes around the world. Attended by hundreds of thousands of protesters, these strikes served to raise awareness and action towards combatting climate change. In addition, the terms *flygskam* (a Swedish term meaning flight shaming) and *tagskryt* (train bragging) movements were born in Sweden, which encouraged travellers to 'clip their wings and swap planes for trains' (Coffey, 2019). As Swedish climate activist Greta Thunberg has shown, refusing to fly as a means of reducing air travel's contribution to climate change can send a powerful signal to politicians and corporations that people are willing to change their behaviour. Indeed, for many of us, a decision not to fly might be the most significant reduction in emissions we can make as individuals (Hendy, 2019), with one of the impacts of Covid-19 being the reduction in pollution from travel related activities, particularly air travel (Henriques, 2020). This encourages us to think about how we can engage in tourism while minimising our environmental footprint, perhaps by using video conferencing for business meetings and virtual reality to satisfy our desire to explore the world. In addition, as a response to the concern about air travel and climate change there has been an associated growth in slow travel (Dickinson and Lumsdon, 2010) and staycations (Molz, 2009).

The potential for 'flight shaming' as a way of influencing consumer behaviour to avoid travel by plane is a concern for island nations who rely heavily on large numbers of tourists flying to and from the country. For example, the Government of Ireland has a target of receiving 11.6 million overseas tourists by 2025 (an increase from 9.5 million in 2018). However, if flight shaming encouraged tourists to change their behaviour this target is unlikely to be met, with one commentator saying in relation to the development of 'flight shaming':

Ours is an island nation. If flight shame really does take off as a mode of consumer behaviour in coming years, how else are the bulk of US and European tourists meant to get here? They won't all suddenly switch to the booze cruise from Holyhead. (Mark, 2019:12).

The impacts of climate change and consumer activism are discussed further in Chapter 10.

Generational cliques

The generation of people that followed the Millenials (born between 1982 and 1996) are known as Generation Z (Dimock, 2019). Contention exists around the years that constitute 'Generation Z', particularly as setting generational cut-off points is not an exact science. Some sources say those born between 1995 and 2010, or 2012 or 2015 constitute Generation Z while others call this generation 'digital natives' or the 'net-gen' (Turner, 2015). In this chapter, Generation Z is defined as individuals born since 1997 to the present day (Dimock, 2019). These people will become the next wave of travellers who will have the time, money and inclination to travel. But how will their travel choices be different from previous generations and what drivers might make their experience different? Will they travel for cultural engagement, meeting people or as a way to learn about themselves? Veal *et al.,* (2001: 415) suggest there are 'at least two contrasting forces at work in leisure choices, first the desire to "do one's own thing", and second the desire to "be together" and do things with other people'. They propose that 'being together' is 'still among the strongest of human drives'. They suggest it seems unlikely people in general will 'forsake social interaction entirely for the sake of technology' (Veal *et al.,* 2001: 415). In addition, they suggest that the 'concept of "fun"' in leisure is very important even when the fun involves 'very basic, simply, social leisure activities'...[which have] 'existed for centuries and which will, no doubt, continue as long as human civilisation exists' (Veal *et al.,* 2001: 410).

Clues towards the preferences of Generation Z are given via a survey reported by Businesswire. The 2019 survey called *The Priceline Generation Travel Index*, surveyed 1,503 demographically representative Americans between the ages of 17 and 65 years old, representing four generations of Americans, namely: Generation Z, Millennials, Generation X and Baby Boomers who reported travelling for vacation at least once in the last year. The sample was asked 'If you were given an extra $100 to spend on vacation what would you spend it on?' Generation Z travellers said they would spend that money on an experience, rather than a nicer hotel or more leg room on a flight. Nearly one in two (48%) Generation Z respondents reported that photos posted to social media inspired them to travel and they were seeking cost effective travel with photo opportunities. They are also reported feeling 'pressure' to 'post the perfect photo' while on vacation – a burden cited by nearly one in three (29%). Ready access to WiFi was particularly important to the group to allow them to 'feed their social channels while vacationing'. The survey showed that Generation Z were the most spontaneous travellers with nearly half (48%) of Generation Z respondents planning their travel within one month of departure. These findings reflect that Generation Z are spontaneous travellers who are concerned about the cost of a holiday, are interested in the holiday experience and expect frequent social media photo opportunities. In contrast to the Generation Z desire to stay connected via social media, in the future there may be an increase in demand for holidays that involve having a 'digital detox'. This is sometimes known as the 'Joy of Missing Out' (JOMO) which is in contrast to the phrase FOMO (Fear of Missing Out). Holidays which reflect JOMO are responding to concerns regarding the impact of constant connectivity in consumers lives, highlight the enjoyment of being mindful of the moment and reflect that some consumers are resisting this 'always-on mentality' (Geerts, 2019: 21).

Collaborative consumption

There has been a recent increase in evidence of *collaborative consumption* around the world, which refers to the sharing, loaning and exchanging of consumer goods (Botsman and Rogers, 2010). The drive for collaborative consumption has occurred due to an increase in environmental concerns, being cost conscious, concern regarding consumerism and a renewed belief in the importance of community which has led to a culture of 'sharing, aggregation, openness, and cooperation' (Botsman and Rogers, 2010: xx). Another term for 'collaborative consumption' could be 'access economy', which has also been used to describe the practice of consumers moving away from ownership and paying to access goods or services for a limited time. Examples from tourism include booking lodging in privately-owned short-term rentals, booking taxis, engaging in a guided tour or dining with locals (Geerts, 2019: 37).

Generation Z is coming of age in this increasingly collaborative world where sharing and collaboration have become second nature and people meet up in chat rooms and social forums to 'upload music, books, and videos; and share thoughts and daily actions with the rest of the world' (Botsman and Rogers, 2010: 86). From a tourism perspective this has led to the establishment of a range of tourism-related opportunities such as the sharing of accommodation, cars, bikes and electric scooters, and the development of traveller reviews and tour guiding to assist with holiday planning and itineraries. Geerts (2019: 37) suggests that travel has been 'revolutionised by the access economy'. Such collaborative consumption suggests less emphasis on physical materialism as a status and perhaps a move to having tourism experiences as status that is confirmed and boasted about via social media posts and photographs.

Consumer activism

As well as a higher level of education and awareness about climate change, recent online exposes and documentaries regarding the use of animals as entertainment and animal ethics may lead to tourists being more mindful to avoid animal use during their tourism experiences (Carr and Broom, 2018). These activities include elephant riding, marine mammal performances and interactions with 'captive dolphins, lion cubs and suspiciously docile tigers' (Cape Times, 2019: 6). They may also be less likely to purchase goods made of ivory, exotic skin products and body parts of endangered animals due to a higher awareness of the ethics of doing so. There may be a growth in philanthropically-minded travellers who show an interest in 'going and doing' projects by volunteering at well-established animal sanctuaries, going on beach clean-ups and on cultural exchanges to engage with locals on a more personal level (Cape Times, 2019: 6).

Summary

This chapter has reflected on the drivers which are likely to influence tourism in the future. Some aspects will remain as drivers of tourism, namely the time, money and means to travel. However, there are likely to be new drivers which influence the future of tourism. Among these may be citizens who are concerned about the impact that air transport has on climate change and who choose to holiday locally without using air travel. Generation Z is likely to continue to travel to seek human interaction and new experiences, but how they travel may be different. They may embrace slow tourism by using land and sea-based transport rather than air transport. Airlines are under pressure to develop aircraft which are fuel-efficient and use renewable sources of energy. A green fleet of aircraft may encourage guilt free travel. Since tourists often seek new cultures and landscapes that are far away from their home environment, it may be impractical not to fly. Instead, tourists will have to choose cleaner alternatives in order to satisfy the human need for new experiences, which is likely to continue driving future tourism for the foreseeable time.

Case study: Analysing the Chinese tourist market

It is expected that the Chinese outbound tourism market will be worth US$365 billion for the 2019-2025 period (PR newswire, 2019). In 2017 the top five destinations for Chinese tourists to visit were Hong Kong, Thailand, Japan, Vietnam, and South Korea, with these countries accounting for over 35% of visitations by the total Chinese outbound tourists (Chow, 2018). The reasons for the growth in the Chinese outbound market reflect a rise in disposable incomes among Chinese citizens and the relaxation of travel restrictions on overseas travel by the Chinese government. The influences on Chinese city residents intention to travel to a foreign country are age, gender, education, annual household income, paid leave days, residential city, an individual's willingness to change life situations, and level of perceived happiness (Wong and Stevens, 2016). Thus, the major factors driving the growth of the Chinese outbound market include a rising affluent and middle-class population, a liberal tourism policy such as Approved Destination Status (ADS), an open-door policy, other relaxed government agreements and an increasing number of passport holders in the country (Chow, 2018).

These societal and economic changes in China have provided the means through which Chinese people can travel overseas and the rapid growth of the Chinese outbound travel over the last 20 years. In the future, China is forecast to overtake the US and Germany to become the largest source of outbound departures by 2030 (Geerts, 2019). For example, the number of Chinese cruise ship passengers more than tripled between 2012 and 2014. China's largest online travel agency named CTrip suggested that the strong demand for cruises among Chinese tourists reflected the Chinese tourists liking an international feeling to the trip but that they also appreciated the reminders of home and the cruise liners were able to cater for large groups, such as big family gatherings (Wong and Stevens, 2016).

The large number of Chinese people who now have the money and ability to travel has a range of implications for many receiving countries, particularly in regard to catering for visitors during popular times of years such as Chinese New Year and the Chinese summer holidays, and in regard to the range and type of accommodation and restaurants available. Chinese tourists want to experience the pleasures of travelling abroad, and may reflect on the extent to which a particular destination can provide 'face' or 'status' when they return home from their adventures in an overseas country. Therefore, Chinese tourists may want to see many exotic attractions which they can photograph, and this may help in their 'bragging rights' on their return home. Chinese visitors may be on a tight schedule but they may want to try different cuisines and shop for gifts to take home, including local products (Hennessy, 2013).

Chinese travellers are increasingly opting for destinations with diverse and exciting offerings, as they explore new ways to experience culture and nature during their holidays, and the demand for adventure travel is also growing. The Chinese tourists are interested in visiting attractions like waterfalls and may be interested in engaging in outdoor and adventure activities like rafting, kayaking and ziplining (Chow, 2019).

Discussion questions

1 Determine the interests and motivations of the Chinese outbound market, and then provide suggestions on how tourism and hospitality businesses in the tourist receiving countries can provide appropriate products and services that encourages repeat business and good word of mouth recommendation.

2 Recognising the economic benefits of the growing Chinese tourist market for receiving countries, suggest what actions governments could take to ensure adequate management of the Chinese market and that any benefits are maximised.

References

Botsman, R. and Rogers, R. (2010) *What's Mine Is Yours: The Rise of Collaborative Consumption*, New York: Harper Collins.

Cape Times (2018) Leave no trace - the new travel mantra, *Cape Times*, 31 December, viewed 12/12/2019, www.pressreader.com/south-africa/cape-times/20181231/textview

Carr, N. and Broom, D. M. (2018) *Tourism and Animal Welfare*, Wallingford: CABI.

Chow, P. (2019) More Chinese travellers hunger after adventure travel, *TTG Asia*, 2 December, viewed 11/12/2019, https://www.ttgasia.com/2019/12/02/more-chinese-travellers-hunger-after-adventure-travel/

Clark, C. (2001) The future of leisure time, in A. Lockwood and S. Medlick (eds.), *Tourism and Hospitality in the 21st century*, Oxford: Butterworth Heinemann, pp. 71-81.

Cocolas, N. Walters, G., Ruhanen, L. and Higham, J. (2020) Air travel attitude functions, *Journal of Sustainable Tourism*, **28** (2), 319-336.

Coffey, H. (2019) When the going gets green, *The Independent*, 30 November, viewed 10/12/2019, https://edition.independent.co.uk/editions/uk.co.independent.issue.301119/data/9219961/index.html

De Botton, A. (2002) *The Art of Travel*, New York: Pantheon.

De Juniac, A. (2019) Opinion: Air Industry is committed to making flying sustainable, *Business Times Singapore*, 12 November, viewed 12/12/2019, www.businesstimes.com.sg/opinion/aviation-industry-is-committed-to-making -flying-sustainable

Dickinson, J. and Lumsdon, L. (2010) *Slow Travel and Tourism*, Abingdon, Oxon: Routledge.

Dimock, M. (2019) Defining generations: Where Millennials end and Generation Z begins. *Pew Research Center*, 17 March, viewed 11/12/2019, www.pewresearch.org/fact-tank/2019/01/17/where-millennials-end-and-generation-z-begins/

Dwyer, L. (2012) Trends underpinning global tourism in the coming decade, in W.F. Theobald (ed.), *Global tourism*, 3rd ed, Burlington, MA: Elsevier Science, pp. 529-545.

Dwyer, L. (2015) 'Globalization of tourism: Drivers and outcomes', *Tourism Recreation Research*, **40** (3), 326-339.

Geerts, W. (2019) *Megatrends Shaping the Future of Travel*, Euromonitor International, viewed 11/12/2019, https://go.euromonitor.com/WTM19.html

Getz, D. (2008) Event tourism: definition, evolution, and research, *Tourism Management*, **29** (3), 403-428.

Hendy, S. (2019) Should you be worried about your carbon footprint every time you take a flight?, *Scroll.in*. 29 November, viewed 1/12/2019, https://scroll.in/article/945069/should-you-be-worried-about-your-carbon-footprint-every-time-you-take-a-flight

Hennessy, B. (2013) My view: Face up to tourist needs, *The Cairns Post*, 27 March, viewed 12/12/2019, www.pressreader.com/australia/the-cairns-post/20130327/textview

Henriques, M. (2020) Will Covid-19 have a lasting impact on the environment?, 27 March, viewed 30/03/2020, https://www.bbc.com/future/article/20200326-covid-19-the-impact-of-coronavirus-on-the-environment

MacCannell, D. (2013) *The Tourist: A New Theory of the Leisure Class*. California: University of California Press.

Mark, P. (2019) Finance: Flight-shaming casts green shadow over tourism, *The Irish Times*, 15 November, viewed 10/12/2019, www.pressreader.com/ireland/the-irish-times/20191115/282883732548285

Molz, J.G. (2009) Representing pace in tourism mobilities: staycations, slow travel and the amazing race, *Journal of Tourism and Cultural Change*, **7** (4), 270-286.

Moltz, J.C. (2011) Technology: Asia's space race, *Nature*, **480** (7376), 171.

Morrison, A. M. (2019) *Marketing and Managing Tourism Destinations*, 2nd edn, Abingdon, Oxon: Routledge.

Page, S. (2019) *Tourism Management*, 6th edn, Abingdon, Oxon: Routledge.

PR Newswire (2019) China outbound tourists visit and spending report 2019 - Forecast to 2025 for the $365 billion industry, *PR Newswire*, 26 September, viewed 13/12/ 2019, https://www.prnewswire.com/news-releases/china-outbound-tourists-visit-and-spending-report-2019---forecast-to-2025-for-the-365-billion-industry-300925281.html

Sharpley, R. (2018) *Tourism, Tourists and Society*, 5th edn, Abingdon, Oxon: Routledge.

Todd, G. (2001) World travel and tourism today, in A. Lockwood and S. Medlick (eds.), *Tourism and Hospitality in the 21st Century*, Oxford: Butterworth Heinemann, pp. 3-17.

Turner, A. (2015) Generation Z: Technology and social interest, *Journal of Individual Psychology*, **71** (2), 103-113.

Veal, A.J., Darcy, S. and Lynch, R. (2001) *Australian Leisure*, 2nd edn, Frenchs Forest: Pearson Higher Education.

Veal, A.J., Darcy, S. and Lynch, R. (2015) *Australian Leisure*, 4th edn, Frenchs Forest: Pearson Higher Education.

Weaver, D. and Lawton, L. (2014) *Tourism Management*, 5th edn, Milton: Wiley.

West, D. M. (2018) *The Future of Work: Robots, AI, and Automation*, Washington DC: Brookings Institution Press.

Whitley A. (2020) How coronavirus will forever change airlines and the way we fly, *Bloomberg,* 25 April, viewed 1/04/2020, www.bloomberg.com/news/features/2020-04-24/coronavirus-travel-covid-19-will-change-airlines-and-how-we-fly

Wong, F. and Stevens, A. (2016) Chinese tourists just can't get enough of cruises, *CNN Wire*, 16 August, viewed 12/12/2019, https://money.cnn.com/2016/08/16/news/companies/china-tourism-cruise-ships/index.html

3 Tourists of the Future

Introduction

Today, more and more people are travelling than ever before, with 1.5 billion international tourist arrivals recorded in 2019 and the forecasted 1.8 billion international arrivals set to be reached well before its predicted 2030 (UNWTO, 2019; 2020). Traditionally, the wealthier industrialised world has predominately been responsible for both the supply and demand of tourism. However, in recent years a gradual shift has occurred with new destinations beginning to challenge these traditional destinations. There is the expectation that 57% of all international tourist arrivals will be in emerging destinations by 2030 (UNWTO, 2017). The rise of the middle class has resulted in more of the world's population gaining access to leisure time and the means to increased international travel. Travel experiences in the past have typically consisted of sun, sand and surf type holidays. Tourists are no longer content with these passive activities, rather seeking more experiential and engaging travel experiences instead. This suggests a change in demand from the mass tourism holidays of the 1970s and 1980s to more individualised tourist experiences (Sharpley, 2018). Drivers of change contributing to these changes in travel demand include increased globalisation along with a variety of economic, social, political, technological and environmental trends (Dwyer *et al.*, 2008). Chapter 2 discusses the key drivers of change, along with several trends considered to have an impact on the future development of the international tourism industry. This chapter explores some of these trends further in the context of future tourist behaviour, namely smart tourism, virtual tourism, smart boredom, super sabbaticals and solo travellers.

Smart tourism

Gretzel *et al.* define tourism as a 'social, cultural and economic phenomenon involving the movement of people to countries or places outside their usual environment for personal or business/professional purposes' (2015: 180). This comes as no surprise to Gretzel *et al.* who see the concept of 'smart' being suitably applied to the tourism phenomena, 'given its information-intensity and the resulting high dependency on information and communication technologies' (2015: 180). Li (2017) observes that smart technology includes technology with a degree of smartness that supports new forms of collaboration and value creation that leads to further innovative, entrepreneurial and competitive changes driven by new technologies. Gretzel *et al.* (2015: 180) specify that smart tourism 'involves multiple components and layers of smartness supported by information and communication technologies (ICTs)' inclusive of hardware, software, groupware, netware and humanware, as the synergies of these systems help:

> …facilitate operational and strategic management of organisations by enabling them to manage their information functions and processes as well as communicate interactively with their stakeholders for achieving their mission and objectives (Neuhofer *et al.*, 2014: 341).

Neuhofer *et al.* add that 'ICTs have become key elements in all operative, structural, strategic and marketing levels to enable interactions among suppliers, intermediaries, and consumers on a global basis' (2014: 342). Smart tourism involves multiple components and layers of smart supported by ICTs inclusive of the destination, tourism experience and local businesses (Gretzel *et al.*, 2015). Lyon and Helsinki were jointly crowned European Capitals of Smart Tourism for 2019 by the European Jury based on the evaluation factors relating to city image, accessibility, sustainable tourism, digitalisation in tourism services as well as cultural heritage and innovativeness in tourism offerings.

At its base, smart tourism refers to 'smart' destinations 'that apply smart principles to urban or rural areas with consideration for both residents and tourists in their efforts to support mobility, resource availability and allocation, sustainability and quality of life/visits' (Gretzel *et al.*, 2015: 180). In addition to the destination component, there is the 'smart' experience component which Gretzel *et al.* defines as one that 'specifically focuses on technology-mediated tourism experiences and their enhancement through personalisation, context-awareness and real time monitoring' (2015: 181). In the years following the development of Web 2.0 and the rise of social media, people have played a more active role in the creation of experiences. An experience rich in meaning results in tourists feeling more empowered, the notion of co-creation builds on this very principle (Neuhofer *et al.*, 2014).

The third component, smart business, refers to the complex network 'that creates and supports the exchange of touristic resources and the co-creation of the tourism experience' (Gretzel *et al.*, 2015: 181). A distinct aspect of this component is the public-private collaboration involving governmental departments adopting a technology-focused approach as providers of data and supporting infrastructure (Gretzel *et al.*, 2015). Importantly, Gretzel *et al.* observe that smart tourism spans three layers across these key components (2015: 181). First, the 'smart information layer that aims at collecting data; a smart exchange layer that supports interconnectivity; and a smart processing layer that is responsible for the analysis, visualisation and intelligent use of data'. In considering these important components, Gretzel *et al.* (2015: 181) therefore define smart tourism as:

> …tourism supported by integrated efforts at a destination to collect and aggregate/harness data derived from physical infrastructure, social connections, government/organisational sources, and human bodies/minds in combination with then use of advanced technologies to transform that data into on-site experiences and business value-propositions with a clear focus on efficiency, sustainability and experience enrichment.

In terms of digital media and adoption of new technology, the arrival of social media has fundamentally changed the nature of business communication with the customer. In today's modern world, and particularly in the realm of tourism, this has reshaped how local businesses and other service providers communicate with the tourist. Bolan and Simone-Charteris define social media as 'a type of media dispersed through online social interactions that can take a variety of forms including social networking sites, blogs, wikis, podcasts, photo and video sharing, social bookmarking and virtual environments' (2018: 730). The modern traveller now avidly utilises social media pre-trip to inform their decision-making, and increasingly uses such platforms during their holiday to enhance their experience and share aspects of their experience with others (Bolan and Simone-Charteris, 2018). Nowadays, tourists' smartphones and tablets receive interpretative and experiential information that signage, leaflets/brochures, guidebooks, maps, and tour guides once provided (Bolan and Simone-Charteris, 2018). In terms of consumption, tourists are no longer passive recipients of information; instead, they are actively engaged in peer-to-peer product recommendations and electronic word-of-mouth (Qiuju and Zhong, 2015).

Such connected experiences via social media and mobile devices provide opportunities for local businesses and other service providers, not only to investment in the growth of their business, but more importantly to facilitate both relevant and meaningful experiential outcomes for tourists and other consumers (Bolan and Simone-Charteris, 2018). There is an expectation amongst tourists regarding the availability of smart facilities. These include smart hotels, smart restaurants, and smart visitor attractions where Wi-Fi is free, of high quality, and always on (Bolan and Simone-Charteris, 2018). Failure to provide high quality Wi-Fi connectivity at event venues, in and around tourism attractions, or in specific hotel and accommodation establishments represents the failure of local businesses and service providers to formulate and facilitate new digital strategies for the future (Paris et al., 2015; Pearce and Gretzel, 2012). The use of beacons, 'a low-cost, micro-location-based technology that use Bluetooth low energy (BLE 4.0) for communicating with beacon-enabled devices in people's smartphones or tablets' (Bolan and Simone-Charteris, 2018: 741), may be distributed across a wide area in facilitating high quality Wi-Fi connectivity. Event organisers of the 2015 US PGA Golf tournament utilised Wi-Fi technology to enable people's mobile devices to receive information beamed or 'pushed' from beacons situated close to spectators. Specifically, event organisers utilised 'live online streaming to introduce new interactive features for mobile devices that included shot-by-shot laser generated data from the course' (Bolan and Simone-Charteris, 2018: 741). To do this involved the installation of beacon technology in the scoreboards carried by volunteers. Beacons received constantly updated data, with remote drones helping also to capture data that spectators could subsequently download via the tournament app on their smartphones (Bolan and Simone-Charteris, 2018: 741).

Similarly, beacon technology may be used in the travel and tourism industry to enhance the tourist experience, from receiving travel information, check in assistance and baggage claim at the airport to guest identification and service provision at a hotel as well as navigation and transportation at a destination (BLE Mobile Apps, 2018). Smart tourism, inclusive of the digital realm of social media, and the important aspect of Wi-Fi connectivity, highlights the growing importance of technology and its influence on the experiential outcome for tourists. Those involved in facilitating tourism experiences need to embrace the digital expectations of tourists and consumers alike. Local businesses and other service providers must actively engage their customers through social media and other digital means, such as virtual reality, if they are to succeed and prosper in the marketplace of the future.

Virtual tourism

In addition to smart tourism, much opportunity exists to revolutionise travel and dining experiences through connection with the latest virtual digital technology (Spence and Piqueras-Fiszman, 2013). Virtual tourism is the application of virtual reality (VR), including augmented reality (AR) and mixed reality (MR), to tourism (Champion, 2019). Whilst VR is popular, interactive and may provide a simulated environment, it does prohibit its user from developing a relationship with the real world due to the immersion in that simulated environment (Kounavis *et al.*, 2012). Meanwhile, augmented reality (AR) enables its user to develop a relationship with their environment due to its ability to superimpose computer-generated data onto the real view and it is for this reason, the technology has increased in recent popularity. According to Mobidev (2019), 'as of May 2019, the installed user base for AR-supporting mobile devices reached 1.5 billion'.

Virtual tourism typically involves reconstructing a virtual version of reality, which eliminates the need to travel to the destination; however, it enhances the overall tourism experience (Champion, 2019). For example, using technology to provide virtual tours of a hotel enables potential guests to 'try before they buy', an opportunity not typically associated with the service industry. VR may facilitate the entire booking process with the user interface experienced through a virtual reality headset, enabling comparisons between hotel rooms and facilities (Revfine, 2020). Virtual travel experiences not only provide great opportunity for prospective tourists to sample some of the main attractions, which ultimately may influence their decision to visit a destination, but also provides the experience for those unable to travel to the destination. It also serves as a sustainable alternative to physical visitation, especially for those sites or destinations suffering from the impacts of over tourism.

Organisations such as the British Broadcasting Corporation (BBC), Google, Netflix and Twitter have produced a number of VR experiences of the Great Barrier Reef, aimed at providing the opportunity to experience the UNESCO site's beauty and wonder but also educating potential tourists on the condition of the reef (Chong, 2017). Google, in conjunction with Californian not for profit CyArk, have posted online 26 world heritage sites in 18 different countries using realistic 3D models created from a collection of digital photography, aerial drones, and a 3D laser scanning technology (Metcalfe, 2018). In addition, VR theme parks exist worldwide designed to add value to destinations including Singapore's HeadRock VR Sentosa, Florida's LEGOLAND Resort 'The Great LEGO Race', Six Flags Magic Mountain's DC Superheroes Drop of Doom VR in California, and the Big Apple Coaster in Las Vegas, to name a few. The world's largest indoor VR Theme Park located in a shopping mall, VR Park, opened in Dubai in 2018 and offers over 25 different experiences (Visit Dubai, 2020).

Figure 3.1: Virtual Reality (VR) Park in Dubai Mall. *Source:* C. Lade, 2020.

Within the hospitality sector, in addition to facilitating menu selection and ordering processes, developments in ICTs in recent times have enabled restaurants to begin incorporating both VR and AR into their restaurant dining experiences (Guttentag, 2010). As dining out is such a sensory experience, it makes sense that operators would start to experiment with how digital technology can be used in a restaurant setting (Tatti, 2016). An example of the use of AR to deliver an imaginative multisensory dining experience includes *Le Petit Chef*, a new and entirely innovative two-hour dining concept whereby one continuous 3D projected story retells the travels of Marco Polo, the Italian merchant, explorer and writer, who travelled through Asia along the Silk Road between 1271 and 1295 (Le Petit Chef, 2020). A six-course menu show transports the diner along a unique culinary journey incorporating a variety of flavours, sights, tastes and sounds.

Another example of the use of AR to deliver a unique multisensory dining experience includes 'The sound of the sea' seafood dish served at *The Fat Duck*, Heston Blumenthal's signature restaurant in Bray, United Kingdom. In addition to their seafood dish, diners receive a seashell containing a pair of iPod earphones with instructions to insert into their ears in order to hear waves crashing softly on a beach, among other sounds of the sea (Spence and Piqueras-Fiszman, 2013). This imaginative dining experience to a degree is a *degustation* menu in which a range of signature dishes are presented with very little or no written description (Frost *et al.*, 2016). The whole experience is referred to as a 'journey' with the menu presented as an itinerary for the day, and diners are advised to allow four hours to complete the 15 dishes that make up the seven courses (The Fat Duck, 2020). According to Moscardo (2010), tourists create stories during their experiences and then present these stories to others as memories of their trip, perhaps a similar notion when creating memorable dining experiences.

Smart boredom and its impact on the travel experience

While the increased utilisation of smart technologies has facilitated the creation of more personalised experiences for the traveller, the constant 'connectivity' aspect may generate a sense of not feeling complete relaxation while on holiday. Smart boredom is considered as the 'use of spare time to browse social media, play games, manage online finances and catch up on the latest news anytime, anywhere!' (Belton, 2018). The notion behind the term 'smart boredom' is maximising our time, whether it be during our morning commute, whilst waiting for an appointment or in the queue at the supermarket. The use of mobile technology via smartphones and tablets along with the ability to connect to the internet, usually via free Wi-Fi, are major facilitators of the smart boredom concept. Of course, there are many advantages of being connected via technology 24/7. We have the potential to become more efficient in our time management, keep our lives up to date, access information in a timely manner, and it enables flexible working arrangements, just to mention a few. However, when it comes to recreation and holidays, the demand for travel experiences reflecting complete relaxation may relate to our ability to be always contactable and online in our everyday life. Tourists of the future may be seeking a 'digital detox' whereby their travel experience may be totally void of any technology and online platforms. Teresa Belton (2018), discusses the need for people to unplug in order to clear their mind, unwind and completely relax. Belton uses the analogy of a farmer learning many years ago that land, allowed to lie fallow from time to time, becomes more productive. China, South Korea, Turkey and South Africa are high users of mobile technology, and therefore there could be future demand for 'digital detox' holiday experiences, particularly from these nations (Belton, 2018). Super sabbaticals, where individuals spend extended periods away from their usual living environment, may provide such an opportunity to do so.

Super sabbaticals

Lyons *et al.* define a gap year as:

> ...a nominal period during which a person delays further education or employment in order to travel. Although people can experience this interlude at any point across the lifespan, it is within the period of early adulthood that the gap year phenomenon became most popular which commonly involved a year off after completing secondary school or tertiary studies (2011: 365).

While a gap year is sometimes seen as consisting of 'full moon parties in Thailand with one hundred US dollars to last two months', overworked billionaires are reinventing this concept so that a gap year is now more like a super sabbatical (Meltzer, 2019). Normally, a sabbatical is an extended period of time during which a person can leave his or her customary work to travel or acquire new skills and/or training. However, a super sabbatical involves:

> ...hard-charging career entrepreneurs, financers, and other 'one-percenters', unaccustomed to half measures at work or play, seeking a maxed-out full throttle gap year to jump start the kind of mind expanding, soul sustaining experiences they missed out on whilst laying the foundations of their start-up company (Meltzer, 2019).

Those taking a super sabbatical are typically aged 35 – 50 years. The common trait is a Type A overachiever, who are either between jobs, have recently sold their company – or in some cases, were experiencing complete work and personal life imbalance (Meltzer, 2019). Reasons why people in need of a break should consider taking a sabbatical or gap year include:

- Seeing destinations, they have always dreamed of visiting;
- Discovering skills, they never knew they had;
- Seeing things prompting them to take action;
- Noticing themes in one's life not previously recognised;
- Opening one's self to big changes; and,
- Realising there is no time limit in trying something new.

Taking a super sabbatical provides opportunity for people to reconnect with themselves, their family and friends as well as the world around them. In some cases, people have an epiphany; they have an eagerness to be in touch with their emotions, to experience those goose bump moments that amaze them. Sometimes, the empathy hits closer to home, as Meltzer (2019) explains:

> …a New York tech founder woke up one day and saw his kids practically towering over him. At that moment, he decided to put his life on hold and focus on his most precious assets. The family, with three kids aged 8 – 15 years, spent eight months in South America. From Antarctica, they worked their way north, trekking through the glaciers of Patagonia, living on an *estrancia* [cattle ranch] in Mendoza, shearing sheep and learning '*gaucho*' rodeo skills, as well as studying Incan culture in Peru and dolphins in Galapagos.

Likewise, Meltzer (2019) points to a London finance executive who negotiated a six-month break between jobs to travel with his wife and young children on a 191-day trip around the world, which included:

> …an astrological reading at the monastery in Bhutan where the royal family have their fortunes told, as well as a trip across New Zealand in a camper van (with a private heli-flight over the Franz Joseph glacier). The family also participated in well digging and crop planting on the Indonesian island of Sumba, where the children joined a class at a local school.

Upon their return, the city banker went back to finance but his wife chose to pursue a career in wellness which she attributed to techniques learned during their trip, whilst the children were 'drawn out of their bubble' (Meltzer, 2019).

Those pursuing volunteer activities as part of a super sabbatical or gap year can undertake holidays that involve projects designed to relieve the poverty of certain groups through community well-being and conservation work. The duration of volunteer projects varies from a few weeks to a month or more (Lyons *et al.*, 2011). These projects may include 'building infrastructure, business development, environmental regeneration, farming, teaching or wildlife conservation' (Lyons *et al.*, 2011: 367). Having a super sabbatical or gap year can result in people becoming more keenly self-aware and connected to family and friends. Not surprisingly, some decide to pursue a career involving philanthropy. Having sought self-development and adventure in a range of unique destinations, people have transitioned from not only just valuing success, but appreciating more so the significance of their work (Meltzer, 2019).

Creativity can also be a key reason why people may take a super sabbatical or gap year. As creative activities are subject to growing time pressure, due to work, family and other personal commitments, there are fewer gaps in people's busy lives for creative activities since they require a heavy time investment.

People who do not have time to engage in creative pastimes in their normal everyday environment are using their leave time for self-development and self-expression. Richards (2002: 2) defines creative tourism as that which 'offers visitors the opportunity to develop their creative potential through active participation in courses and learning experiences that are characteristic of the holiday destination where they are undertaken'. The demand is for active participation, rather than passive modes of engagement, as part of a desire for learning and personal development (Richards, 2002). Whilst on a gap year or super sabbatical, people can participate in a wide range of activities including arts and crafts, design, gastronomy, health and healing, languages, spiritualty, nature and sports. Personal engagement (or connection) by people is generally agreed to be at the heart of all experiential outcomes. These types of experiences not only rely upon achieving something fulfilling, unique or special, but also that which is educational and transformational (Pine and Gilmore, 1998; 1999).

Equally, it can be something that enhances an individual both socially and culturally, as well as that which is challenging, participative or re-affirming (Prentice, 2004: Ooi, 2005). Richards (2002) suggests the field of gastronomy offers a range of different experiential products including wine tasting tours, gastronomic holidays and cooking academies, with cuisine being an ideal way for people to learn about food, its link to local culture and agriculture, as well as learning to cook local dishes whilst taking a super sabbatical or gap year. Individuals are seeing the benefits of taking a sabbatical or gap year to develop themselves or reassess their career prospects. Very often, people taking a long break from work may need to show some result from their trip after returning home, either in terms of personal fulfilment or through the acquisition of new skills to either their employer or family members. For them, the time spent travelling was no longer an end in itself, but the means to a new future. The opportunity to engage in a super sabbatical presents opportunities for individuals to pursue solo travel experiences as a way of personal growth and development.

The solo traveller

Solo travelling is on the rise. In 2018, 70% of Intrepid Travels USA trips involved solo travellers (Intrepid Travel, 2018). Key motivations for solo travel include for relaxation and time to unwind, meeting new people, redefining your comfort zone, to increase confidence, independence and self-sufficiency and ultimately, seeking 'freedom'. Solo travellers are typically aged between 41 and 47 years, have an income in the $150,000 range and 85% are women (Solo Traveller World, 2019). According to a 2017 Princeton Survey Research

Associates study, 58% of millennials worldwide are prepared to travel solo compared to 47% of older generations, with 26% of millennial women having already travelled solo (Schroeder, 2019). Solo travellers are inclined to be highly or well educated, want to see more of the world and not willing to 'wait for others' in order to do so, considered frequent travellers with approximately 46% of respondents travelling three or more times a year, and generally seeking adventure type experiences over urban getaways (Solo Traveller World, 2018). A number of travel companies are catering for this increase in solo travel and particularly, solo women travellers. Intrepid Travel recorded a 40% increase in solo passengers on their group trips in the five years to 2018 and have launched their solo only tour range to Bali, India and Vietnam as well as waiving single supplements and creating communal tables for shared meals (Shabada, 2018). Saga Cruises (UK) has also incorporated 109 new solo cabins on its 'Spirit Discovery' ship launched in 2019 to accommodate this increased demand for such travel. More recently, there has been the publication of a number of guidebooks to accommodate this growing segment of the travel market (Figure 3.2).

Figure 3.2: Lonely Planet's *The Solo Travel Handbook. Source:* C. Lade, 2020.

Summary

Smart technology makes it possible for tourists to participate in the development or customisation of their own experiences. Smart tourism involves tourists using their smart phone or tablet to access web-based information at a smart destination that enhances and adds value to their experience (Gretzel *et al.*, 2015). Gen Z travellers particularly expect digital connectivity and frequent social media opportunities; they tend to be spontaneous, be concerned about the cost of travel and more focused overall on experiences (Priceline Generation Travel Index, 2019). Destinations and services are experimenting with the use of virtual and augmented reality in order to enhance the tourist visitor experience, while the growing trend of 'smart boredom' needs to be acknowledged in conjunction with changing consumer demand and increased technology expectations. Super sabbatical trips, regarded as an extension of the Gap year traditionally undertaken by school leavers, are an opportunity for middle aged career driven individuals to engage in travel experiences they may have missed out on while too busy building their careers. The number of people travelling worldwide is continually increasing and their travel needs are shifting in response to a number of key drivers of change, including globalisation. The various trends identified in this chapter relate to a range of economic, social, political, technological and environmental factors and the tourism industry must recognise the influence of these factors as they evolve in order to meet traveller needs and remain competitive.

Case study: Personalising the travel experience

With an estimated 86% of travellers valuing personalisation of their experiences, a number of major international airlines and hotel brands are developing ways to implement personalisation (Gilliland, 2017). Opportunities for inflight personalisation via mobile technology and guest service tools along with the use of geo-targeting, which involves sending specific offers to potential customers based on location, exist in order to enhance the customer experience. Incorporation of VR technology may also provide travellers with the opportunity to 'try before you buy', providing the ability to experience hotels, airplanes and tourist destinations before purchase and ultimately, play a role in better meeting consumer needs and therefore contribute to their overall tourist experience.

KLM Royal Dutch Airlines utilise extensive personalisation via their digital marketing techniques. The airline sends personalised emails to retarget those customers that have abandoned online carts, enabling them to resume their user online journey. iFly50 is an anniversary edition of the airline's brand magazine, which incorporates stunning imagery and an interactive user experience where customers select their five favourite destinations for a chance to win a trip of a lifetime; the airline may then follow up with additional targeted marketing activities in relation to their selected favourite destination. Furthermore, the airline introduced a 'meet and seat' feature on their intercontinental flights which integrates social networking enabling their passengers with a reservation to link their Facebook or LinkedIn profiles with their flight. This may enable passengers to connect before, during or following their flight.

Delta Airlines is also implementing features aimed at personalising the inflight experience having recently soft launched its guest service too. The handheld device enables flight attendants to access detailed passenger information inclusive of their frequent flyers status, specific dietary requirements and/or special needs assistance. Singapore Airlines have also introduced options for customised meals, with the airline allowing all passengers travelling in suites, first, business and premium economy class to choose their main course from the in-flight menu via the airline's website or mobile app any time between three weeks and 24-hours prior to flight departure (Chen, 2019). Parents travelling with children in all cabins can also view and pre-order specific child meals on flights departing Singapore up to 24 hours before departure.

Best Western Hotels and Resorts wanted to increase the number of downloads of its mobile application and was able to do so by identifying the device recipients used to open the email and alter its message accordingly (Gilliland, 2017). People using an Apple device receive directions to the Apple app store, while Android users receive directions to the Google Play store. This streamlined the process and encouraged users to download the app without having to locate the correct link themselves. The hotel group also used geo-targeting to send specific and relevant offers based on the recipient's location (Gilliland, 2017). Depending where the recipient was when they opened the email, they received different recommended destinations. Both these digital strategies proved to be effective, with the hotel chain experiencing a 143% increase in downloads of its app compared to similar campaigns. Similarly, there was a 10% increase in email click-through rates by non-rewards members (Gilliland, 2017). Virgin Hotels Chicago uses mobile technology to enhance the customer's experience during their stay by allowing guests to customise their hotel experience via their existing mobile device. A mobile app called 'Lucy' enables guests to personalise their stay by adjusting the room temperature, streaming content via their hotel TVs and making external dining reservations (Gilliland, 2017). Virgin Hotels also launched its new preference program 'The Know', replacing the usual hotel loyalty programs, where guests may complete an online questionnaire requesting certain services and disclosing specific information such as mini bar selections, newspaper preferences and allergies. Such a program allows hotel brands to compete with the likes of Airbnb concerning the more personal, intimate touches and subsequently, guests are more likely to return in the future (Gilliland, 2017).

Discussion questions

1 Identify factors contributing to the increased demand for more personalised and customised tourist experiences.
2 What challenges may exist in the delivery of more personalised services and experiences by the provider?
3 When you travel, do you expect to receive personalised service? If so, why?

References

Belton, T. (2018) Why boredom can be good for you, *The Conversation*, 23 January, viewed 15/01/2020, theconversation.com/why-boredom-can-be-good-for-you-90429

BLE Mobile Apps (2018) *Use of Beacon Technology in Travel and Tourism Industry*, 13 June, viewed 5/02/2020, https://www.blemobileapps.com/blog/use-beacon-technology-travel-tourism-industry/

Bolan, P. and Simone-Charteris, M. (2018) Shining a digital light on the dark: Harnessing online media to improve the dark tourism experience, in P.R Stone, R. Hartmann, T. Seaton, R. Sharpley and L. White (eds.) *The Palgrave Handbook of Dark Tourism Studies*, London: Palgrave Macmillan, pp. 727-746.

Champion, E.M. (2019) Virtual reality adds to tourism through touch, smell and real people's experiences, *The Conversation*, 5 March, viewed 3/02/2020, https://theconversation.com/virtual-reality-adds-to-tourism-through-touch-smell-and-real-peoples-experiences-101528

Chen, J. (2019) Singapore Airlines extends meal pre-ordering to premium economy, *Business Traveller Asia-Pacific*, 26 March, viewed 24/01/2020, https://www.businesstraveller.com/airlines/2019/03/26/singapore-airlines-to-offer-more-personalised-and-diversified-in-flight-dining-options/

Chong, Z. (2017). Dive Australia's Great Barrier Reef with Netflix and Google, *CNET*, 25 October, viewed 4/02/2020, https://www.cnet.com/news/dive-great-barrier-reef-vr-virtual-reality-with-netflix-google/

Dwyer, L., Edwards, D.C., Mistilis, N., Roman, C., Scott, N. and Cooper, C., (2008) *Megatrends underpinning tourism to 2020: analysis of key drivers for change*. Gold Coast, Queensland: CRC for Sustainable Tourism.

Fat Duck. (2020) viewed 29/01/2020, http://www.thefatduck.co.uk

Frost, W., Laing, J., Best, G., Williams, K., Strickland, P. and Lade, C. (2016) *Tourism, Gastronomy and the Media*, Bristol: Channel View Publications.

Gilliland, N. (2017) How six travel & hospitality brands use personalisation to enhance the customer experience, *Econsultancy*, 26 June, viewed 1/02/2020, https://econsultancy.com/how-six-travel-hospitality-brands-use-personalisation-to-enhance-the-customer-experience/

Gretzel, U., Sigala, M., Zhang, X. and Koo, C. (2015) Smart tourism: Foundations and developments, *Electronic Markets*, **25** (3), 179-188.

Guttentag, D. (2010) Virtual reality: Applications and implications for tourism, *Tourism Management*, **31** (5), 637-651.

Intrepid Travel. (2018) Ten Travel Trends For 2018, viewed 10/03/2020, https://www.intrepidtravel.com/travel-trends-2018/

Kounavis, C., Kasimati, A. and Zamani, E. (2012) Enhancing the Tourism Experience through Mobile Augmented Reality: Challenges and Prospects, *International Journal of Engineering Business Management*, **4** (10), 1-6.

Le Petit Chef. (2020) viewed 29/01/2020, http://www.lepetitchef.com/

Li, Y. (2017) Individuals' motivations to adopt smart technologies for tourism - discrepancy between initial and post adoption, in Streitz N., Markopoulos P. (eds) *Distributed, Ambient and Pervasive Interactions. DAPI 2017*. Lecture Notes in Computer Science, vol 10291, Cham: Springer, pp.77-92, viewed 1/08/2019, https://link.springer.com/chapter/10.1007/978-3-319-58697-7_6

Lyons, K., Hanley, J., Wearing, S. and Neil, J. (2011) Gap year volunteer tourism: myths of global citizenship, *Annals of Tourism Research*, **30** (1), 361-378.

Meltzer, M. (2019) The latest travel trend? Super sabbaticals, *Conde Nast Traveller*, n.d., viewed 10/06/2019, www.cntraveller.com/article/the-latest-travel-trend-super-sabbaticals

Metcalfe, T. (2018) Now you can visit world heritage sites in virtual reality *NBC News Digital*, 22 April, viewed 5/02/2020, https://www.nbcnews.com/mach/science/now-you-can-visit-world-heritage-sites-virtual-reality-ncna867881

Mobidev. (2020) 9 augmented reality trends to watch in 2020, *Mobidev.biz*, 30 June, viewed 1/04/2020, https://mobidev.biz/blog/augmented-reality-future-trends-2018-2020#:~:text=As%20of%20May%202019%2C%20the,the%20pace%20of%20industry%20growth

Moscardo, G. M. (2010) The shaping of tourist experience: The importance of stories and themes, in M. Morgan, P. Lugosi, and J. R. B. Ritchie (eds), *The Tourism and Leisure Experience. Consumer and Managerial Perspectives*, Bristol: Channel View, pp. 43–58.

Neuhofer, B., Buhalis, D. and Ladkin, A. (2014) A typology of technology-enhanced tourism experiences, *International Journal of Tourism Research*, **16** (4), 340-350.

Ooi, C.S. (2005) A theory of tourism experiences: The management of attention in T. O'Dell and P. Billing (eds.) *Experiencescapes: Tourism, Culture and Economy*, Copenhagen: Copenhagen Business School Press, pp. 51-68.

Paris, C.M., Berger, E.A., Rubin, S. and Casson, M. (2015) Disconnected and unplugged: experiences of technology induced anxieties and tensions while traveling, in Tussyadiah I., Inversini A. (eds), *Information and Communication Technologies in Tourism,* Cham: Springer, viewed 10 June 2019, https://link.springer.com/chapter/10.1007/978-3-319-14343-9_58

Pearce, P. and Gretzel, U. (2012) Tourism in technology dead zones: documenting experiential dimensions, *International Journal of Tourism Sciences,* **12** (2), 1-20.

Pine, B.J. and Gilmore, J.H. (1998) Welcome to the Experience Economy, *Harvard Business Review,* **76** (4), 97-105.

Pine, B.J. and Gilmore, J.H. (1999) *The Experience Economy: Work is Theatre and Every Business a Stage,* Boston: Harvard Business School Press.

Prentice, R.C. (2004) Tourist familiarity and imagery, *Annals of Tourism Research,* **31** (4), 923-945.

Priceline Generation Travel Index. (2019) Generation Z Takes Flight, *Business Wire,* 24 September, viewed 5/01/2020, https://www.businesswire.com/news/home/20190924005207/en/

Qiuju, L. and Zhong, D. (2015) Using social network analysis to explain communication characteristics of travel-related electronic word-of-mouth on social networking sites, *Tourism Management,* **46** (C), 274-282.

Revfine, (2020) How virtual reality is transforming the travel industry, viewed 4/02/2020, https://www.revfine.com/virtual-reality-travel-industry/

Richards, G. (2002) Creating a new tourism? in Garcia, S. (ed.) *Turisme I Cultura. Debats Del Congres de Turisme Cultural SITC,* Barcelona: Fundacio Interarts, viewed 15/08/2019, https://independent.academia.edu/gregrichards

Schroeder, B. (2019) Baby Boomers, Millennials and Gen Z are all changing the $8 trillion travel industry in the same way, opening up major opportunities for entrepreneurs and marketers' *Forbes,* 13 November, viewed 15/01/2020, https://www.forbes.com/sites/bernhardschroeder/2019/11/13/baby-boomers-millennials-and-gen-z-are-all-changing-the-8-trillion-dollar-travel-industry-in-the-same-way-major-opportunities-for-entrepreneurs-and-marketers/#7d11601e279c

Shabada, L. (2018) 10 Travel Trends for 2018: Travel's most buzzworthy destinations, emerging trends and sought-after trips, viewed 8/8/2020, https://www.intrepidtravel.com/travel-trends-2018/

Sharpley, R. (2005) The tsunami and tourism: A comment, *Current Issues in Tourism,* **8**(4), 344-349.

Solo Traveller World, (2019) Solo Travel Statistics and Data: 2019 – 2020 viewed 1/02/2020, https://solotravelerworld.com/about/solo-travel-statistics-data/

Spence, C. and Piqueras-Fiszman, B. (2013) Technology at the dining table, *Flavour,* **2**(16), 1-13.

Tatti, E. (2016) Check out how virtual reality is being used in these restaurants, *The Typsy Blog*, 31 May, viewed 10 May 2020, http://blog.typsy.com/check-out-how-virtual-reality-is-being-used-in-restaurants

UNWTO. (2017) *International Tourism Highlights: 2017 Edition*, UNWTO, viewed 1/12/2019, https://www.e-unwto.org/doi/pdf/10.18111/9789284419029

UNWTO. (2019) *International Tourism Highlights, 2019 Edition*, United Nations World Tourism Organisation, Madrid, viewed 10/11/2019, https://doi.org/10.18111/9789284421152

UNWTO. (2020) International tourism growth continues to outpace the global economy, January 2020, viewed 15/07/2020, https://www.unwto.org/international-tourism-growth-continues-to-outpace-the-economy

Visit Dubai. (2020) viewed 4/02/2020, https://www.visitdubai.com/en/pois/vr-park

Additional resources

Le Petit Chef – Marco Polo: https://www.youtube.com/watch?v=KZ3JbvZMzWA

Le Petit Chef – Dessert: https://www.youtube.com/watch?v=LXyX-OvZlUg

Open Heritage Project (Google and CyArk): https://www.youtube.com/watch?v=V-0pJ2-ELUA

4 Hospitality of the Future

Introduction

The topic 'Hospitality of the Future' examines how rapidly the hospitality sector is changing and highlights innovative products and services that are becoming available. For example, space hotels, underwater accommodation, cruise ships and the use of robots are becoming more common in the hospitality sector. Although some of these innovations have not come to fruition, discussions regarding embracing new products and services should be undertaken rather than rejecting ideas based on previous industry expectations. In addition to the introduction of new products and services, technological change challenges traditional employment models of hospitality workers with a real concern regarding employment opportunities in the future that is also explored in Chapter 7. This chapter highlights some of the future aspects of hospitality services with a focus on the rise of robots in the Asian hospitality sector. A further analysis of the advantages and disadvantages of using artificial intelligence (AI) is given. The chapter concludes by suggesting new and changing technology will have an impact on the hospitality industry, however the extent of impact will differ for each type of hospitality business. Case studies are included to illustrate themes that reflect current customer service practices and potentially what the future holds.

Hospitality of the future overview

There is emerging research into the future of hospitality and the direction it will take based on new and changing technologies. First, the traditional function of the hospitality sector will always exist. This is due to the natural need for food, water and sleep. The physiological requirement for sustenance is essential for all creatures to survive. As Calpaldi stated 'eating is arguably the most fundamental of human activities' (1996: 1). Additionally, sustenance does not necessarily need consumption in a restaurant or other hospitality venues, however, humans crave different foods, tastes and experiences. A restaurant for example, can satisfy these desires. Second, water is another element for the function of life and survival (Saltmarch, 2001). In terms of hospitality, this can extend to other beverages; soft drinks, cocktails, wines and beers among others, nonetheless all hospitality establishments will have water available as a staple beverage. Sleep is the third essential human function considered in the hospitality sector. All humans must sleep at some point to function properly (Horne, 1988). Hence, humans generally need a designated place to sleep. When people travel, they require safe, comfortable and clean accommodation facilities. Different standards of accommodation depend on location, cost, service standards, safety and product quality. It is therefore logical that food, water and sleep will always be required as a minimum for human existence. Other factors that may be considered can be loosely based on Maslow's Hierarchy of Needs (1943), including psychological, safety, belonging, esteem and self-actualisation concerns. Chung-Herrera (2007) suggests that successful businesses in hospitality attempt to satisfy all these levels from a customers' viewpoint especially psychological and safety needs. Guests are more likely to stay longer and become repeat customers if the overall experience surpasses their expectations.

Future of space travel for tourists

Although some of the discussions in this chapter have not actually occurred, it is important to keep an open mind about the possibility of ideas coming to fruition in the future. First, space travel is not currently a commercial reality, however it will be someday. The 'space race' as it is known, highlights the desire to become the first commercial entity in space that is not heavily subsidised by government (Seedhouse, 2010). Although many governments invest in a space program mainly for communication or military aspirations, there has been a shift to a commercial model and countries such as China, Japan and North Korea have economic aspirations as does America and Russia (Moltz, 2011). For space travel to become commercially viable, it is essential that paying customers' needs and wants are satisfied, which is a different focus to previous government policies.

Travelling into space for any length of time will require food, beverages and inevitably sleep; the same for humans on Earth. However, the consumption of these essential activities will not be the same. Strickland (2012) outlines the differences between hotels on Earth and potential hotels in space. His research suggests that initially space hotels will be small, cramped and require extensive guest training before venturing into space. This is most likely due to cost restraints and technological challenges. What is certain is the first space hotels will not be the same as space vessels depicted in science-fiction movies. Challenges such as overcoming gravity, food production and water collection are real concerns let alone being able to travel in hyperspace (which is quicker than the speed of light), as seen in the *Star Wars* and *Star Trek*' movie franchises.

Strickland (2017) also considered the skills required by the first space hotel employees. Traditionally a good hospitality worker has four main attributes: 1) good work ethic, 2) being punctual, 3) being responsible, and 4) always accountable (Alonso and O'Neill, 2011). Tasks of a food and beverage attendant on Earth generally include taking orders, carrying plates and glasses, pouring drinks and polishing cutlery for example. Strickland (2017) suggested that these labour-intensive tasks are not required for the first space hotel employees. Instead, the skills required may be how to use equipment to heat food, overcome challenges of consuming a hot beverage, ability to use a toilet, application of telecommunications, administering first aid and perhaps pilot a space shuttle that are considered more vital. Already, there are fundamental differences in the skills required by hospitality workers simply based on the emerging reality of space travel.

Emergence of more underwater hotels

Similarly, underwater hotels have materialised although only a limited number of establishments are currently available. At present, underwater hotels do require the traditional attributes of a hospitality worker to fulfil guest expectations. Conrad Maldives' *The Muraka* on Rangali Island is an example of an underwater hotel. Visiting the website, you will notice that it has the same offerings of a five-star rated hotel. However, some hotel rooms and the restaurant are submerged. The underwater rooms are in shallow water and therefore not highly pressurised, allowing guests to roam freely in comfort (Conrad Maldives Rangali Island, 2020). As this is an exclusive hotel and resort, the price is high in comparison to other hotels; however, the experience may still be affordable and attractive to many. The quality of the facilities and services are comparable to many other hotels that offer underwater hotel rooms. For instance, other 5-star rated underwater hotels appear in Dubai, Florida, Sweden, Maldives, Zanzibar, St Lucia and Fiji, all with a high room rate and service standards.

Increase in cruise ship popularity

Cruise ships are not a new phenomenon however they are becoming more technologically advanced and customer service focused, which is an innovation in itself. In fact, the cruise ship industry has been one of the fastest growing sectors in tourism (Gmelch and Kaul, 2018). A large American based company titled Global Ocean Cruises (GOC) offers cruise vacations globally. Global Ocean Cruises' largest customer base is American with eleven and a half million passengers cruising annually. Figure 4.1 indicates the number of passengers over the last decade for GOC, which mirrors the increasing trend of the overall global cruise ship industry.

Global Ocean Cruises: Passengers (Millions)

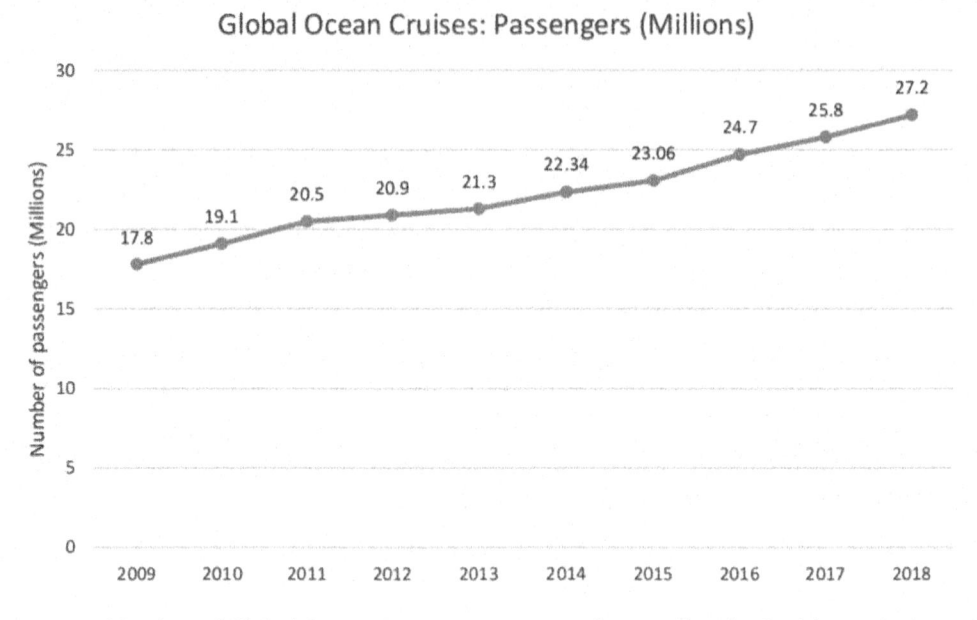

Year	Passengers (Millions)
2009	17.8
2010	19.1
2011	20.5
2012	20.9
2013	21.3
2014	22.34
2015	23.06
2016	24.7
2017	25.8
2018	27.2

Figure 4.1: Number of Global Ocean Cruises passengers. *Source*: Florida-Caribbean Cruise Association, 2018.

Furthermore, GOC states that the most popular sailing routes have been throughout the Caribbean (35%), followed by the Mediterranean (15%), other ports of Europe (15%), China (12%), Australia/New Zealand (7%), Asia (7%), Alaska (5%) and South America (4%) (Florida-Caribbean Cruise Association, 2018: 1). These statistics showcase the most important cruise ship markets and where potential growth can occur. For example, Africa and Antarctica are not destinations for GOC, but this may change in the future to increase capacity and destination appeal.

To assist with service quality and product offerings, GOC conducts exit polls of passengers regarding the attraction of cruise ships. The most popular responses with a 75% or above satisfaction rating were 'relaxing and getting away from it all', 'being hassle-free' and 'offers something for everyone' (Florida-Caribbean Cruise Association, 2018: 6). All-inclusive packages include food and non-alcoholic beverages, accommodation, on-board activities and entertainment. Additional costs that guests have to pay include alcoholic drinks, gambling money and on shore activities (when the ship is in port).

The Florida-Caribbean Cruise Association (2018) has anticipated emerging markets, that include Southeast Asia and South America, will assist in expanding the cruise ship market further. As standards of living rise in these regions, travel will become more affordable, more people have a willingness to travel and have a higher disposable income. The demand for cruises will dictate how large the cruise ship industry will expand, and Asia may benefit by accepting more cruises and higher tourist numbers. The expectation had been that the traditional cruise ship markets such as Europe and the United States of America would continue to increase with growing populations, and Asia becoming just as attractive to visit regularly.

However, with the global impact of Covid-19 everything has changed. Since March 2020, the cruise ship industry has been decimated. While updating this chapter the pandemic is in the embryonic stage, and hence, the future of the cruise ship industry is unclear. Speculative commentary has suggested that the cruise ship industry will not recover if the global shutdown is for longer than six months. Others have suggested that the whole industry will be overhauled as cruise ships were turned away from ports to return to their place of origin. A complicating factor is that many cruise ships reside in tax havens such as Barbados and no-one is accountable. Mass tourism at ports, environmental concerns, poor wages and cramped working conditions for cruise ship staff have also been exposed, placing a negative light on the industry. As of August 2020, cruise ships continue to take bookings for 2021 indicating a return to normality, but will it be the same? Only time will tell.

Customer service of the future in the hospitality sector

All sectors of the hospitality industry strive for exceptional customer service. In addition to the four main attributes of a hospitality worker mentioned, exceptional customer service also includes empathy, warmth, competence, pleasant personality, honesty, efficiency, punctuality, a willingness to learn, undertake given tasks and an awareness of personal hygiene and grooming (Bufquin *et al.*, 2017). Exceptional customer service not only instils a sense of pride in the individual, it also assists in the profitability of the business. Studies have highlighted that if restaurant staff display desirable characteristics, customers are more likely to order more or become a repeat customer, adding to the revenues of the establishment (Wu and Liang, 2009). Similarly, if restaurant staff members are enjoying their workplace, they are more likely to work harder, longer and be less likely to resign (AlBattat and Som, 2013). Furthermore, financial rewards may flow through gratuities, tips, promotions and increased compensation for workers, which can also be a competitive advantage.

If delivering exceptional customer service is something to strive for and many hospitality worker attributes require human emotions, why would the hospitality industry contemplate substituting human labour with electronic robots? As technology advances at a rapid rate and becomes more and more acceptable in daily life, it is logical for the next generation of hospitality to include robots. Broadbent (2017) hypothesised that microprocessors and computing will allow robots to enter the workforce in great numbers, whereas Gladstone (2016) has stated that artificial intelligence, robots and service automation are effectively being adopted by the travel and hospitality sectors and this is particularly evident in Asia.

Robots in the hospitality sector in Asia

In the words of Ivanov *et al.*, 'robots have arrived and are here to stay' (2017: 1513). Some argue that they are not replacing humans, but they are simply changing the tasks that humans do (Osawa *et al.*, 2017). The most common robots featuring in hospitality today undertake manual tasks such as cleaning, delivering room service, aiding the elderly or as a substitute for human companionship. This is due to existing technology allowing robots to undertake these types of tasks. Essentially, robots are a 'relatively autonomous physical device capable of motion and forming a service' (Murphy *et al.*, 2017: 106). Service automation and robots 'provide vast opportunities to travel, tourism and hospitality companies to improve their operations and productivity, deliver consistent product quality and transfer some of the service delivery process to the customers' (Ivanov *et al.*, 2017: 1502). This is why the hospitality sector has a willingness to embrace robots. These authors also suggest that the most common form of technology used in the hospitality sector is service automation. For example, in restaurants, this includes robotic chefs, baristas, servers, dishwashers, automated ordering, entertainment and payment systems (Ivanov *et al.*, 2017).

Kim and Qu (2014) highlight that some hotels have automated check-in kiosks whereas Berezina (2015) and Trejos (2015) suggest mobile devices are increasingly utilised to boost the speed of information delivery, convenience and enhance the customer experience. Asian hotels have cases of robots already implemented in front office stations, concierge desks, guest relation areas, housekeeping (vacuuming), porterage and room service for increased efficiency (Rajesh, 2015). Figure 4.2 showcases a robot that delivers room service to hotel rooms in Singapore.

Figure 4.1: Jeno from Hotel Jen in Singapore delivering room service. *Source:* Paul Strickland, 2020.

There is emerging research on the main benefits of robots to the service sector in general based on three overarching factors: affordability, adoption potential and overall use of the robot (Murphy *et al.*, 2017).

1. Affordability refers to the overall costs associated with designing, manufacturing, implementing and maintaining a robot. Costs vary depending on the materials used, efficiencies in manufacturing, economies of scale, installation, software and technology required for the robot to function and robot maintenance. This is often the main concern, especially for smaller hospitality operators.

2. Questions need to be asked when deciding the practicality of implementing a robot. Hospitality enterprises would ask questions such as 'Can a robot replace a hotel worker? A restaurant worker? A travel agent? If so, where would the robot be located? Where would the robot be stored when not in use? Does our business have the technological infrastructure to accommodate one or many robots? Do we have the maintenance skills for a robot?' These questions would all help establish the adoption potential of robots.

3. Overall use of the robot related to the actual tasks as 'the advances in robotics and artificial intelligence, increased robot capabilities coupled with decreased purchase and maintenance costs will make robots a viable alternative to human employees in travel, tourism and hospitality companies' (Ivanov *et al.*, 2017: 1512). The question is: 'How will robots achieve these perceived benefits?'

To understand the current use and benefits of robots, Ivanov discusses the term '*robonomics*' which 'is an economic system that uses robots, artificial intelligence and (service) automation technologies as production factors, instead of human labour' (2017: 2). Originally, robonomics focused only on the financial benefits of robots (Crews, 2016), now the acceptance of robots is added. An article published in 2017 titled 'Robots in Hospitality: Five Trends on the Horizon' (see additional resources in this chapter) highlight the following five concepts:

1. Robots assist humans with mundane tasks
2. Robots will create more jobs
3. Hospitality will lead the way for embracing robots
4. Robots and data collection
5. Acceptance of robots and removal of anxiety

1. Robots assist humans with mundane tasks

Robots can perform manual tasks that many employees are not wanting or willing to do. These include tasks such as cleaning, vacuuming, carrying heavy items and repetitive tasks in packaging, labelling and general manufacturing. When robots do not influence actual tasks such as these, robots become a more accepted alternative to human labour. Countries such as China, South Korea and Japan have already embraced robots in hotels and restaurants to undertake these tasks. The Henn na Hotel in Japan is an ideal example. This is the world's first fully automated and robotic hotel. From check-in, to dining in the restaurant, automation is everywhere. The hotel uses facial recognition for your room access and is totally secure. The hotel uses cameras to monitor, although there is a staff member on duty 24 hours a day to monitor the hotel and unforeseen emergencies such as medical or fire events (Henn na Hotel, 2020).

2. Robots will create more jobs

This is particularly true for the travel and hospitality sectors of the tourism industry. By having a higher demand for services due to better efficiencies, occupancy levels in hotels will rise, restaurants will have a higher table turnover and people will have a better travel experience. As the robotics sector grows, there will be an increased demand for robot designers, programmers, engineers, maintenance crews and technology professionals. In response to these emerging occupations, educational institutions will need to create programs to meet demand and offer further qualification in the field of robotics. Robot sales positions for instance, will be required to demonstrate the capabilities of robots that will also need to be upgraded in the future, generating more sales. However, Bowen and Morosan (2018) acknowledge that traditional human-to-human service models will continue to exist, especially in less technologically advanced countries where wages are comparatively low. Singapore, Japan, South Korea and Hong Kong are the exception to this rule in Asia, as these countries have a higher potential to invest in technology that a vast number of people can afford.

3. Hospitality will lead the way for embracing robots

Bowen and Morosan (2018) highlighted a current labour shortage in the service sector and predict that robots will fill this void. If robots become a common occurrence, the hospitality sector should be creating additional roles for humans. This may include developing new and creative service models to offer value and improved service standards for guests. As robots are relatively new and still in the embryonic stage, the potential to conceptualise new ideas and advancements is endless. Hence, it is an opportunity for the hospitality sector to be at the forefront of artificial intelligence and service automation amelioration.

In contrast, the manufacturing industry where utility and safety is paramount, the aesthetic appearance of service automation is not a major concern for guests because the majority do not see this machinery. Conversely, robots in the hospitality sector will be front of house, their construction and appearance will need to be completed in such a way that is acceptable to customers. These robots should be approachable, non-threatening, convenient and helpful. To overcome the fears of customers, robots should display attributes that are more human-like for guests to feel comfortable. When designing the robot, roboticists should incorporate characteristics such as empathy and willingness to assist whilst having movement, shape and sounds that put the guest at ease. Having robots that reflect more human-like characteristics may achieve this goal and be better suited to the hospitality service sector.

Ho and MacDorman (2010) postulated that animated robots can easily replicate human emotions; however, Yu (2019) argues that if robots are too human-like and of a perceived high intelligence, it makes guests feel uneasy, scared and disturbed. Therefore, a balance needs to be achieved. To illustrate these findings, the Henn na Hotel in Japan has both a human-like robot on the front office desk and a dinosaur dressed in a porter cap (bellhop) and bow tie. This hotel is trying to appease all guests by offering different front desk options that all guests feel comfortable using (see Henn na Hotel, 2020).

4. Robots and data collection

A celebrated academic named Robert Boyce once said 'knowledge is power' (Boyce, n.d.). In the hospitality sector, that refers to information about the guests, their expectations and experiences. Robots can play an important role in gathering data and information based on these factors. Robots can store data to increase the customer experience, especially during future transactions. For example, a customer in a restaurant may order a beer before the meal. Then, on a repeat visit, the robot may use facial recognition technology to identify the same customer and suggest the same beer as previously ordered. 'Connie the Concierge' at Hilton hotels is another example of a robot adding to the customer experience and hopefully leaving a positive impression (Kuo *et al.,* 2017). Although customer profiling and collection of data of guests can be a positive for both the company and guest, it may not always be so. Other customers may feel uncomfortable regarding how much personal information is being gathered. Tung and Au (2018) highlighted that robots which gather too much information may make the customer feel uneasy or develop a sense of an invasion of privacy. These authors argued that collecting too much data may put the customer off, especially if they do not know where personal information is being stored and who has access. This opens the debate of the legalities of data collection by robots, how it is stored and used. Legislation by governments will need to address these concerns if this is going to become common practice.

5. Acceptance of robots and removal of anxiety

For some people, robots spark a concern for safety and they become very anxious. Having a real fear of robot intelligence, capabilities and ability to deliver accurate information or services is legitimate, argue Tussyadiah and Park (2018). There is also evidence that the majority of robots do not display human attributes such as empathy, therefore today's robots are more task orientated than simulating desirable attributes of a human hospitality worker. Nonetheless, this is slowly changing. There are service robots designed to be companions for humans and display more human and communicative characteristics. In fact, Ivanov *et al.* suggest 'service robots are designed to support and service humans through physical and social interactions' (2017: 1502). These tend to be companion robots and some humans are already receptive. When humans become more tolerant and accept robots, it is envisaged that fear and anxiety regarding robots will be lessened if not removed entirely.

Advantages of robots

There are many advantages to using robots in hospitality. Ivanov (2017) has stated there is an economic benefit. At present, it is relatively inexpensive to purchase a robot when compared to paying staff wages. Robots can range from $US 6,500 – $20,000 and perform tasks to which an employer would have to pay a traditional worker a salary of more than US$50,000 annually. Second, robots can work 24-hours a day making them more efficient. Robots require no sick or family days or bonuses, and there is no need for training or hiring or being let go. Third, robots can perform tasks to set service standards, be programmed for a variety of different duties, have their capabilities expanded with hardware and software updates and improve quality of the customer experience in a timely manner (Papanyan, 2017). Fourth, robots do not convey a negative attitude, drink alcohol, smoke cigarettes or take illicit drugs. Additionally, 'robots do not complain, get ill, go on strikes, spread rumours, discriminate, quit their job without notice, show negative emotions, shirk from work' but do undertake routine and undesirable tasks (Ivanov, 2017: 3). Finally, it will become more important that robots speak different languages which is very desirable in the tourism industry (Wilcock *et al.*, 2017).

Disadvantages of robots

In general terms, when there are advantages, typically disadvantages can be also be identified. The most obvious concern is mass job loss for hospitality workers (Ivanov, 2017). Although this is a real threat for some specific occupations, the literature has indicated that robotics will create more jobs in hospitality such as in maintenance, software development and hardware requirements for robots. This means traditional skills will be transferred to other areas that support the robotic industry, such as a traditional kitchen hand learning how to maintain an automated commercial dishwasher.

There are also suggestions that robots reduce the 'human touch' element of the service industry (Pieska *et al.*, 2013). This is due to the hospitality sector not knowing exactly how much to invest. To illustrate, if the hospitality sector was to replace all workers with robots and society rejects the technology because of perceived fear and anxiety, the investment is wasted. However, robots currently require human supervision therefore total adoption of robots will be gradual. Additionally, robots require guidance and structure, and lack creativity, therefore requiring human intervention for quite some time.

Summary

This chapter discussed the concept of 'hospitality of the future'. The discussion featured the potential future of space hotels, current offerings of underwater hotels, trends in the cruise industry and the use of robots to replace hospitality workers. This chapter suggested that tourists will always require food, beverages and sleep therefore there will always be a need for the hospitality sector. The quality and service standards may vary from country to country, hotel type and style, and these considerations will always depend on the price guests are willing to pay. Technology is advancing at a rapid rate and the hospitality sector is starting to embrace the potential benefits by automating some routine and manual jobs. The economic benefits of robots is attractive for businesses coupled with other cost savings that may help to explain the implementation of robotics. The disadvantages highlighted the maintenance side of robots, lack of human emotion and customer anxiety.

This chapter adds to the body of literature in the hospitality sector and future outcomes. By presenting international examples of the future of hospitality, it attempts to fulfil the gap in the research providing useful case studies based on current practices. This research was limited to Asia, however, could feature other international locations such as Europe and North America. The number of case studies offered does not cover everything that is currently available in hospitality and robot literature. Future studies should include continued economic benefits of robots, customer expectations and perceptions of the adoption of robots, ethical concerns, health and safety concerns and the impact on human employees. The case study from Asia will demonstrate the implementation of robots in the hospitality sector.

Case study: The benefits of robots in a restaurant in Asia

To illustrate the benefits of robots in Asia, please review the video clip that can be located at the URL below. Students are to watch the video clip and then answer the questions below. Students should examine the acceptance of robots in Asia, compare this to other international locations and discuss the legislation for the robots actions.

View the clip https://www.youtube.com/watch?v=6ap5bFYF-yw

Discussion questions

1   Robots already feature in restaurants in Southeast Asia. Explain if the replication of this phenomenon is possible within other continents based on technology, price, clientele and acceptance.

2   In the future, is there the possibility that robots will be capable of handling quality control, health and safety standards of restaurants, or will this have to remain a human task due to an individual country's legislation? If a robot were to accidentally give a customer food poisoning, who is responsible? Explain your thoughts.

References

AlBattat, A.R. and Som, A. (2013) Employee dissatisfaction and turnover crises in the Malaysian hospitality industry, *International Journal of Business and Management*, **8** (5), 62-71.

Alonso, A.D. and O'Neill, M.A. (2011) What defines the 'ideal' hospitality employee? A college town case, *International Journal of Hospitality & Tourism Administration*, **12** (1), 73-93.

Berezina, K. (2015) Mobility convergence. Hotel business review, *Hotel Executive*, 4 January, viewed 17 May 2019, http://hotelexecutive.com/business_review/4178/mobility-convergence

Bowen, J. and Morosan, C. (2018) Beware hospitality industry: the robots are coming, *Worldwide Hospitality and Tourism Themes*, **10** (6), 726-733.

Boyce. R. (n.d.). Robert Boyce Quotes, *AZQuotes.com*, viewed 4/02/2020, https://www.azquotes.com/author/37616-Robert_Boyce

Broadbent, E. (2017) Interactions with robots: The truths we reveal about ourselves, *Annual Review of Psychology*, **68** (1), 627-652.

Bufquin, D., DiPietro, R., Orlowski, M. and Partlow, C. (2017) The influence of restaurant co-workers' perceived warmth and competence on employees' turnover intentions: The mediating role of job attitudes, *International Journal of Hospitality Management*, **60**, 13-22.

Capaldi, E.D. (1996), *Why We Eat What We Eat: The psychology of eating*. Washington, DC: American Psychological Association.

Chung-Herrera, B.G. (2007) Customers' psychological needs in different service industries, *Journal of Services Marketing*, **21** (4), 263-269.

Conrad Maldives Rangali Island (2019), viewed 15/06/2019, https://www.conradmaldives.com/

Crews, J. (2016), *Robonomics: Prepare today for the jobless economy of tomorrow*. New York: CreateSpace Independent Publishing Platform.

Florida-Caribbean Cruise Association (2018) *2018 Cruise Industry Overview*, viewed 30/01/2020, https://www.f-cca.com/downloads/2018-Cruise-Industry-Overview-and-Statistics.pdf

Gladstone, N. (2016) Are robots the future of hotels?, *Oyster*, 13 June, viewed 15/06/2019, https://www.oyster.com/articles/are-robots-the-future-of-hotels/

Gmelch, S.B. and Kaul, A. (2018) *Tourists and Tourism: A reader*, 3rd edn, Long Grove, IL: Waveland Press.

Henn na Hotel (2019), viewed 15/06/2019, https://www.h-n-h.jp/en/

Ho, C.C. and MacDorman, K.F. (2010) Revisiting the uncanny valley theory: Developing and validating an alternative to the Godspeed indices, *Computers in Human Behaviour*, **26** (6), 1508–1518.

Horne, J. (1988) *Why We Sleep: The functions of sleep in humans and other mammals*. Oxford; New York: Oxford University Press.

Ivanov, S. (2017) Robonomics - principles, benefits, challenges, solutions, *Yearbook of Varna University of Management,* **10**, 283-293.

Ivanov, S.H., Webster, C. and Berezina, K. (2017) Adoption of robots and service automation by tourism and hospitality companies, *Revista Turismo and Desenvolvimento,* **27** (28), 1501-1517.

Kim, M. and Qu, H. (2014) Travellers' behavioural intention toward hotel self-service kiosks usage, *International Journal of Contemporary Hospitality Management,* **26** (2), 225-245.

Kuo, C.-M., Chen, L.-C. and Tseng, C.-Y. (2017) Investigating an innovative service with hospitality robots, *International Journal of Contemporary Hospitality Management,* **29** (5), 1305-1321.

Maslow, A.H. (1943) A theory of human motivation, *Psychological Review,* **50** (4), 370-396.

Moltz, J.C. (2011) Technology: Asia's space race, *Nature,* **480** (7376), 171.

Murphy, J., Hofacker, C. and Gretzel, U. (2017) Dawning of the age of robots in hospitality and tourism: Challenges for teaching and research, in *European Journal of Tourism Research,* **15**, 104-111.

Osawa, H., Ema, A., Hattori, H., Akiya, N., Kanzaki, N., Kubo, A., Koyama, T. and Ichise, R. (2017), What is real risk and benefit on work with robots?: From the analysis of a robot hotel, in *Proceedings of the Companion of the 2017 ACM/IEEE International Conference on Human-Robot Interaction*, New York: Association for Computing Machinery, pp. 241-242.

Papanyan, S. (2017) When robots do it all and leisure is mandatory: Not for another 100 years, *US Economic Watch,* 22 June 2017, BBVA Research, 1-8.

Pieska, S., Luimula, M., Jauhiainen, J. and Spiz, V. (2013) Social service robots in wellness and restaurant applications, *Journal of Communication and Computer,* **10** (1), 116-123.

Rajesh, M. (2015) Inside Japan's first robot-staffed hotel, *The Guardian,* 14 August, viewed 3/03/2019, https://www.theguardian.com/travel/2015/aug/14/japan-henn-na-hotel-staffed-by-robots

Saltmarsh, M. (2001) Thirst: or, why do people drink?, *Nutrition Bulletin,* **26** (1), 53-58.

Seedhouse, E. (2010) *The New Space Race: China vs. USA,* Chichester: Praxis Publishing.

Strickland, P. (2012) Do space hotels differ from hotels on earth? The mystery is solved, *Journal of Hospitality Marketing & Management,* **21** (8), 897-908.

Strickland, P. (2017) The first space hotel employees: Human resources challenges in a post-terrestrial paradigm, *Journal of Human Resources in Hospitality & Tourism,* **16** (4), 445-458.

Trejos, N. (2015) Marriott to hotel guests: We're app your service, *USA Today*, 13 May, viewed 11/04/2019, http://www.usatoday.com/story/travel/2015/05/13/marriott-hotels-mobilerequests-two-way-chat/27255025/

Tung, V.W.S. and Au, N. (2018) Exploring customer experiences with robotics in hospitality, *International Journal of Contemporary Hospitality Management*, **30** (7), 2680-2697.

Tussyadiah I.P. and Park, S. (2018) Consumer evaluation of hotel service robots, in B. Stangl and Pesonen, J. (eds.), *Information and Communication Technologies in Tourism 2018 Proceedings of the International Conference in Jonkoping, Sweden, January 24-26, 2018*, Cham: Springer, pp.308-320.

Wilcock, G., Laxstrom, N., Leinonen, J., Smi,t P., Kurimo, M. and Jokinen, K. (2017) Towards SamiTalk: A Sami-speaking robot linked to Sami Wikipedia, in K. Jokinen and G. Wilcock (eds.), *Dialogues with Social Robots.* Lecture Notes in Electrical Engineering, 427, Singapore: Springer, pp. 343-351.

Wu, C.H.J. and Liang, R.D. (2009) Effect of experiential value on customer satisfaction with service encounters in luxury-hotel restaurants, *International Journal of Hospitality Management*, **28** (4), 586-593.

Yu, C-E. (2019) Humanlike robots as employees in the hotel industry: Thematic content analysis of online reviews, *Journal of Hospitality Marketing & Management*, **29** (1), 23-28.

Additional resources

Connie the Concierge: https://newsroom.hilton.com/corporate/news/hilton-and-ibm-pilot-connie-the-worlds-first-watsonenabled-hotel-concierge

'Robots in Hospitality: Five Trends on the Horizon' highlight the following concepts: https://hospitalitytech.com/robots-hospitality-five-trends-horizon

5 The Future of Visitor Attractions

Introduction

This chapter explores the future of visitor attractions by using the examples of theme parks and world-renowned tourist attractions to illustrate the possible developments in this area. The chapter considers the impact of social media on visitor attractions and the associated phenomenon of 'overtourism' to demonstrate the types of problems being faced by visitor attractions and reflect on how tourism authorities are addressing these issues.

Theme parks and other visitor attractions

Attractions are arguably the most important component in the tourism system and could be described as the main pull factor for tourists to travel to particular destinations (Dann, 1981). Since attractions are the core of the tourism product, if there were no attractions then tourism as we know it would not exist (Swarbrooke, 2012). While no clear definition exists of visitor attractions, they can be split into four main types, namely:

- Features within the natural environment, such as rainforests, waterfalls, beaches, lakes and rivers;
- Human-made buildings, structures and sites that were designed for a purpose other than attracting visitors, but which now attract substantial numbers of visitors who use them as leisure amenities, e.g., places of religious worship such as cathedrals and temples;
- Human-made buildings, structures and sites that are designed to attract visitors and are purpose-built to accommodate their needs, such as theme parks and amusement parks; and,
- Special events (Swarbrooke, 2012).

A popular type of purpose-built attraction is the theme park. When a visitor buys a ticket and passes through the turnstiles, this signals they are entering a different space. This space offers a variety of rides, restaurants, shops, and shows that are all themed around one or several past, exotic, or fictional cultures which are 'geographically, visually, and ritually separated from the rest of the world' (Freitag, 2017: 706). Theme parks have different admission policies that include, pay-as-you-go and pay-one-price, but all visitors have the opportunity to spend on food, souvenirs and other related purchases. In the future, theme parks are expected to increase in number due to the growth in urban population, the rise in the middle-class population with associated increase in disposable income and, an increase in international tourism expenditure (Business Wire, 2019).

Figure 5.0: Walt Disney Resort, Orlando, Florida. *Source:* L. McCartney, 2020.

Probably the best-known theme park operator is Walt Disney which operates Walt Disney World Resort in Orlando Florida that encompasses Magic Kingdom, Animal Kingdom, Epcot and Hollywood Studios. Walt Disney World Resort is a key player in Orlando's $75.2 billion tourism industry as it attracts 75 million visitors annually, accounting for 70% of the region's tourism market share, with these visitors spending money at local hotels, restaurants and shops (Bilbao, 2019). In addition, the Disney Company operates Disneyland Resort, California; Hong Kong Disneyland Resort; Shanghai Disney Resort, China; Tokyo Disney Resort; and, Disneyland Paris. As with all successful theme parks, Disney is continually introducing new rides utilising new technology as a way of investing in its theme parks. Disney is also interested in achieving maximum return from its intellectual properties. As a result, a new ride centred on the latest Star Wars release called the Star Wars: Rise of the Resistance ride, based at the Hollywood Studios theme park in Orlando, opened in December 2019 to correspond with the associated movie release. The ride consists of a 15-minute experience, with pre-show and post-show elements to create one story. This may help to minimise the feeling by visitors of standing in a very long and time-consuming line to enter the ride (Bilbao, 2019).

Other theme parks have reported the introduction of facial recognition technology to replace paper or online tickets. For example, the new Universal Studios theme park under construction in China will use facial recognition for admissions, payments and queueing. The facial recognition cameras will be linked to an electronic payment platform, with the new technology allowing visitors to use their faces to join express queues for rides, pay for meals and open lockers (MacDonald, 2019). Since waiting in line is one of the aspects of a theme park which is not popular among tourists, the introduction of virtual queuing would go some way to make the visitor experience more positive. Torres *et al.* (2018) note that queues have been an endemic problem in the theme park and attraction industry and theme parks have tried to alleviate waiting by creating virtual queues, improving queue design, and developing interactive experiences while waiting.

Research has shown that visitors to theme parks have the potential to be 'delighted' with their experience and the aspects most likely to create a positive experience being positive value perceptions, limited waiting times in queues for the rides, an excellent core product, quality food and beverages, servicescape, pricing decisions and sensible admissions policies (Torres *et al.*, 2018). On the other hand, guests who are 'outraged' with their experience describe various aspects of their visit to the theme park which create negative experiences such as the perceptions of value, long waiting periods in queues, poor customer service, low quality or deficient core products, poor quality of food and beverage, poor facility maintenance, aggressive pricing decisions, poor staff selection, training, and working conditions or aggressive admissions policies (Torres *et al.*, 2018).

In the future, there is likely to continue to be demand for attractions that encourage engagement between different members of the tourist travel party and allow and encourage playful behaviour. For example, there may be a growth in escape room attractions which provides interactive group games that require the visitors to solve puzzles as a way of finding their way out of a locked area (Bleiberg, 2019) Escape rooms may be particularly popular among families and friends travelling together as it encourages them to engage with each other, solve some problems and be 'immersed in an experience' with the use of live actors to make the escape challenge all the more realistic. These escape room attractions are expected to become more hi-tech with a move away from the basic structure of solving clues to opening of locks and boxes with keys as well as being expected to grow in number at tourism destinations and on cruise ships (Escape Mate, 2018).

Theme parks, rides and roller-coaster sometimes take a dark and scary approach to enticing consumers. In the future it is possible that the market for death and horror-based attractions and entertainment will continue to grow and it is predicted that there will be the development of 'terror parks' in the future which may allow tourists to revisit the tragedies of the recent past. This is already seen in the growth of dark tourism in response to such high-profile sites such as the TV series *Chernobyl* which has triggered tourists to visit the nuclear disaster site with the need to use respirators and dosimeters to avoid radiation exposure. Other dark sites and attractions in Europe which are expected to continue to be popular for tourists include, the Anne Frank House in Amsterdam, the Berlin Wall in Germany, the Auschwitz concentration camp in Poland and the Somme Battlefields in France (Wright, 2018).

The bucket list, selfies and social media

The concept of ticking off items on a bucket list occurs when an individual wants to complete a list of experiences and/or achievements before they die. This reflects the notion that travel experiences offer self-fulfilment and are a measure of the success or meaningfulness of one's life (Thurnell-Read, 2017). Crossing off an item on one's bucket list may be similar to the collection of an artefact, with bucket list items traditionally being rare and only being able to be achieved after 'significant planning and effort' (Smith, 2019: 13). However, in recent years the bucket list concept has become a vehicle for 'culturally specific ideas of what constitute(s) "good" tourism experiences' (Thurnell-Read, 2017: 65). In other words, there is a change in the concept of the bucket list which suggests that tourism needs to be 'done' in the right way, and that individuals are now encouraged to 'desire a constantly renewing range of tourism experiences' (Thurnell-Read, 2017: 58). From a visitor attraction perspective, bucket list tourists may visit well-known tourist attractions. However, they may only spend a small amount of time at the site due to the main purpose of visit being purely to 'tick' off the attraction from their list.

The importance of posting one's travel experience on social media appears to have led to an increase in the number of people wanting to visit well-known attractions and then posting a selfie photograph from that location. For the photograph to be suitable for posting on social media individuals want themselves to be in the frame which has led to the desire to take selfies at well-known attractions, often using the selfie stick to allow the attraction to be in the background with the individual in the foreground. Macintyre (2019) suggests that in some tourism cultures, the selfie-in-front-of-the object is 'virtually obligatory' with Facebook and Instagram creating a culture that requires the traveller to be pictured in a place which others will 'recognise and admire'. The rise of the bucket list mentality means there is a 'social premium' on specific destinations which may lead to thousands of tourists converging on the same place, to 'take photographs of themselves alongside others doing the same thing' (Macintyre, 2019). Scott (2016) reported on a survey which found that nearly half of the British tourists interviewed took selfies at an iconic landmark and then left just minutes later, suggesting the tourists were more interested in taking selfies for social media to share with friends and family, rather than exploring the sites. In addition, 25% of those interviewed admitted they have travelled somewhere for the sole purpose of taking a photograph to post on social media, with almost one in ten tourists (8%) uploading their first holiday picture within an hour of arriving, and 11% sharing within the first 12 hours of the holiday.

Kugel (2016) suggested that 'we shouldn't go to places because they're world-famous; we should go to fully appreciate the thing that made them world-famous, for example to experience an unparalleled collection of Renaissance art'. Similarly, Mahdawi (2019) believes that travel today is less about broadening a person's horizons and more about building a personal brand. To illustrate the point, Mahdawi (2019) used the example of people visiting a concrete flight of stairs located in the Bronx, New York where the actor Joaquin Phoenix, dressed as a clown, danced down the stairs in the film *Joker*. That scene later became a meme. She suggests that people want to visit an attraction or location to recreate the image they have viewed on screen or online and then use the associated hashtag (in this case #jokerstairs) to allow their selfie of the site to be viewed and shared. She suggests this development reflects a new term named 'meme tourism', where people travel to recreate famous scenes (and often humorous memes).

Unfortunately, the taking of selfies has resulted in the accidental deaths of at least 259 people since 2011 (Bansal *et al.*, 2018). Such deaths may reflect that tourists are searching for more and more dramatic ways to create self-portraits for social media. These extreme selfies are leading people into more ridiculous and dangerous antics that pre-social-media they would never have dreamt of attempting (Parnell, 2017). Deaths by selfies has resulted in some 'no-selfie zones' in tourist areas, especially in places with bodies of water, mountain peaks, tall buildings, bottle necks and religious gatherings prone to lethal stampedes. This no-selfie rule is designed to decrease the incidence of selfie-related deaths, due to this 'exponential' increase in the number of people dying while trying to photograph themselves (Matthews-King, 2018). In addition, selfie-sticks are now banned in many museums and parks, including Walt Disney Resort. The Walt Disney Company said that using a selfie stick is interrupting the flow of visitors and impacting on the enjoyment of others (Weigold, 2018). In addition, the 'selfie-in-front-of-the object' tends to slow down the movement of crowds and interferes with the enjoyment of others. One solution would be to prohibit photography at peak periods, or restrict photography to specific hours, bookable at a higher cost. There have even been suggestions that wide conveyor belts could be installed in front of the most overcrowded exhibits and tourist attractions to prevent loitering, inhibit photography, and keep queues flowing (Macintyre, 2019).

Tourism behaviour associated with bucket lists, selfies and social media raises the question of whether in the future this is likely to continue? The answer to this is to reflect on human nature and to think about why tourists are interested in selfies today and if this might continue in the future. Weigold (2018) asks: what is it about the self-portrait that's so resonant as a form of communication? Also why, psychologically, might someone feel so compelled to snap the perfect selfie that they would risk their life, or the lives of others? Social psychologist Leon Festinger (1954) developed the social comparison theory which proposed that people have an innate drive to evaluate themselves in comparison with others to improve how we feel about ourselves (self-enhancement); evaluate ourselves (self-evaluation); prove we really are the way we think we are (self-verification); and, become better than we are (self-improvement). Selfies may be useful to support our self-worth, particularly because selfies attract more online attention and more comments than any other photographs, and our friends and peers reinforce selfie-taking by giving 'likes' and other forms of online approval (Weigold, 2018). In addition, Weigold (2018) suggests that the approval which comes from 'likes' and positive comments on social media is rewarding, particularly for the lonely, isolated or insecure. He notes that those most likely to post selfies appear to have lower self-esteem than those who do not. Thus, he suggests that posting selfies may help to boost an individual's confidence by showing others how 'awesome' they are; offer proof of an exciting life; expressing an individual's mood; help preserve important memories; share important experiences; demonstrate the individual's unique experiences and show personal beauty and attractiveness (Weigold, 2018).

Instagram and overtourism

Instagram, an image-based platform, uses hashtags as a way of 'creating connections and community by tagging images with keywords that will appeal to niche demographics' (Swan, 2019: 42). When Instagram users take photographs of tourist attractions and destinations, it can inspire tens of thousands of Instagrammers to try to capture the exact same image (O'Connor, 2019) and this has intensified the desire of travellers to show they visit locations with celebrity status (Financial Times, 2019). While the internet might have been expected to introduce world travellers to new and dispersed destinations, it appears to be helping to funnel them into a smaller number of must-see hotspots (Macintyre, 2019). In response to this growth in desire to capture the perfect photograph for Instagram, some tourist operators have begun to cater to this demand. For example, the UK has developed an Insta-friendly map for islands in the Scottish Hebrides to help tourists seeking the best shots of their holidays for their social feeds (Ross, 2019). Hotel designers are now briefed to include Insta-friendly features, such as an eye-catching mural, a hot tub with a view, an infinity pool with colour-changing lights, an aquarium bar, minimalist white decor and sunset views to ensure the location is irresistibly photogenic (O'Connor, 2019). The images not only help to make the destination look more attractive but also may lead to the destination being visited due to providing photo opportunities (Carter, 2018). In the future these Instagram-influenced tourism choices are likely to continue, with influencers and bloggers being brought to attractions by tourist authorities to experience, photograph and share the destination and attractions with their followers. Often photographs on Instagram show a lone tourist depicted on a vantage point above an otherwise empty landscape displaying an 'array of aspirational experiences ranging from seeing the Bagan sunrise to skydiving to posing with a camel by the Egyptian pyramids' (Smith, 2019: 13). However, the reality is that a few steps away there are hundreds of tourists waiting to take exactly the same photo as a means of providing over eight million people on Instagram with content for the tag #bucketlist.

Although there are many thousands of tourist attractions around the world, there are a lesser number of attractions which are described as 'must-see'. As a result, the must-see attractions have become very popular and are now suffering from over-visitation, known as 'overtourism', which has been exacerbated by the demand to take selfies at these well-known sites (Table 5.0 lists the best-known world tourism sites). The term 'overtourism' applies where local people or tourists feel that a place is visited by too many tourists and that this changes its character, causing the location to lose authenticity. The term is new to the tourism literature, but the concept is not and is used to describe the consequences of tourism in some destinations (Capocchi *et al.*, 2019).

Table 5.0: The world's most popular top 10 tourist cities listed alphabetically

Amsterdam, Netherlands
Barcelona, Spain
Berlin, Germany
Copenhagen, Denmark
Dubrovnik, Croatia
Kyoto, Japan
New York City, United States
Paris, France
Reykjavik, Iceland
Venice, Italy

Source: Groundwater, 2018.

Groundwater (2018) highlighted the problem of overtourism by saying:

> There are just so many of us. There have never been more people travelling than there are right now. We're everywhere, but most particularly we're in the places that have always been popular: in European hotspots such as Barcelona, Amsterdam and Venice, in ancient capitals such as Kyoto, and in modern-day hubs such as New York. Travellers are pouring into these places in their millions, in unprecedented numbers, and the destinations can't cope.

Similarly, Cooke (2019) suggests there are too many people who want to travel and the 'famous cities, tropical islands and coastal towns can't accommodate them all'. Macintyre (2019: 30) suggests that world tourism is reaching 'peak capacity, with consequences that include chronic overcrowding, damage to fragile sites, cultural and physical disruption for local people and a degraded experience for the tourists themselves'. In addition, the locals complain of noise pollution, 'crowded parks, pressure on public facilities and rising rents, with the Airbnb revolution warping house prices' (Macintyre, 2019).

Concern about the impact of tourism on cultural and natural sites and a desire to protect them has led to over 1,100 sites in some of the world's most visited countries being placed on the UNESCO World Heritage Site listing. These sites are of immense cultural, religious and historic significance (Table 5.1 lists all the World Heritage listed sites in Australia and a sample from Italy).

Table 5.1: UNESCO Heritage List of cultural and natural sites in Italy (20/55) and Australia (20/20)

Italy	Australia
Historic centre of Rome	Great Barrier Reef
Historic centre of Florence	Kakadu National Park
Piazza del Duomo, Pisa	Willandra Lakes Region
Venice and its lagoon	Lord Howe Island Group
Ferrara, City of the Renaissance	Tasmanian Wilderness
Historic centre of Naples	Gondwana rainforests of Australia
Historic centre of Siena	Uluru-Kata Tjuta National Park
Historic centre of the city of Pienza	Wet Tropics of Queensland
Archaeological areas of Pompei, Herculaneum and Torre Annunziata	Shark Bay, Western Australia
Botanical Garden (Orto Botanico), Padua	Fraser Island
Portovenere, Cinque Terre, and the islands (Palmaria, Tino and Tinetto)	Australian fossil mammal sites (Riversleigh / Naracoorte)
Assisi, the Basilica of San Francesco and other Franciscan Sites	Heard and McDonald Islands
City of Verona	Macquarie Island
Isole Eolie (Aeolian Islands)	Greater Blue Mountains area
Villa d'Este, Tivoli	Purnululu National Park
Genoa: Le Strade Nuove and the system of the Palazzi dei Rolli	Royal Exhibition building and Carlton Gardens
The Dolomites	Sydney Opera House
Medici villas and gardens in Tuscany	Australian convict sites
Mount Etna	Ningaloo coast
Vineyard landscape of Piedmont: Langhe-Roero and Monferrato	Budj Bim cultural landscape

Source: UNESCO, 2019.

In addition, tourism authorities have discussed measures to reduce overtourism. Perhaps tourists could be encouraged to discover 'more places that are off the beaten path, to lessen overcrowding at major tourist hotspots'

(Tahseen, 2019), to recognise that the 'world is a big place, and there's a lot more to explore than the Instagram-famous spots' (PR Newswire, 2019). Can tourists be encouraged to spread out to visit other areas to reduce the focus on a limited number of sites? Can they be persuaded that instead of visiting the obvious places, 'the standard exhibitions, the icons such as the Mona Lisa that are already pictured on a million postcards, there is more to discover, and more enjoyment to be found (and perhaps even more "likes" to garner on social media), in the less obvious destination or artwork' (Macintyre, 2019)? As a result of overtourism, tourism authorities in the industry are encouraging people to travel out of the peak season, to go to the fringes of cities and to encourage people to engage in unique travel experiences (PR Newswire, 2019). A panel of travel bloggers suggested a list of 30 emerging and alternative destinations around the world as a way to encourage more tourists to move away from these tourist hotspots. Such destinations include Georgia, Rwanda, Gothenburg, Sweden, Azerbaijan, The Cook Islands, among others (Table 5.2).

Table 5.2: Top 30 emerging destinations in the world listed alphabetically

Aarhus, Denmark	Launceston, Tasmania, Australia
Azerbaijan	Lausanne, Switzerland
Churchill, Manitoba, USA	Lebanon
Colorado Springs, Colorado, USA	McLaren Vale, South Australia
Cook Islands	Namibia
Dominica	Navarre Beach, Florida
Ecuador	Nova Scotia, Canada
Estonia	Northern Territory, Australia
Ethiopia	Paraguay
Georgia	Paso Robles, California, USA
Gothenburg, Sweden	Poznan, Poland
Kyrgyzstan	Prince Edward Island, Canada
Laos (Southern area)	Rovinj, Croatia
Laurel Highlands, Pennsylvania	Rwanda
La Paz, Mexico	Uzbekistan

Source: Travel Lemming, 2019.

Another idea for handling overtourism is to develop dedicated resorts, which are built to withstand a large number of tourists to the site; the argument being that if tourists are siphoned into dedicated holiday spaces, then this may provide valuable relief to destinations suffering under the weight of their popularity. Recent developments suggest the possibilities of new private island resort 'mega theme parks' which are 'isolated worlds created for the sole purpose of accommodating tourists, many with huge brand names and budgets behind them' (Cooke, 2019). For example, similar to already existing all-inclusive resorts, Disney recently announced it will be creating such a destination in the Bahamas as a stop on its Disney Cruise Line. By keeping tourists in tourist-only areas designed exclusively for them, tourists can have an unrestricted experience while staying out of the way of the locals, and this creates the potential to protect overrun of destinations and attractions (Cooke, 2019). Tourism authorities around the world have also had to devise ways to minimise the impact of overtourism and have introduced various measures such as limiting the number of people per day allowed to enter certain areas of a city. This is likely to continue in the future with more and more destinations and attractions introducing such measures as a way of protecting these popular sites. Table 5.3 lists some responses to overtourism in a range of cities around the world.

Table 5.3: City response to overtourism

Barcelona, Spain	The construction of new hotels in the centre prohibited
Hong Kong	To promote lesser known attractions such as the beach and hiking
Machu Picchu, Peru	A ticket system introduced to limit the number of visitors allowed in high season to 5,000 per day.
Edinburgh, Scotland	Introduction of a tourism tax on hotel rooms
Santorini, Greece	Introduction of a maximum number of 8,000 visitors per day
Maya Bay, Phi Phi Lee, Thailand	Area closed to tourism until 2021 due to corals being destroyed by pollution
Bruges, Belgium	Limited number of cruise ships to town from five to two per day
Komodo Island, Indonesia	Closed to tourism during 2020 due to poachers and pollution from excess visitors
Venice, Italy	Introduced an entry fee due to pollution, flooding and gentrification
Taj Mahal, India	Limit of three hours per person at the site plus an extension of visiting hours to spread out the number of visitors
Amsterdam, Netherlands	Locals renting their homes via Airbnb restricted to only 30 days a year.
Dubrovnik, Croatia	Two cruise ships permitted to dock a day. A limit of the number of visitors allowed to enter the old city

Source: Macau News Agency, 2019; CE Noticias Financieras English, 2019.

Summary

To operate a successful visitor attraction in the future it is important to ensure that the customer is provided with a well-designed site which uses technology to enhance the visitor experience; which provides content and locations to appeal to the visitors' desire for content for their social media presence. Recognition that people are interested in engaging with each other suggests that providing ways for visitors to interact with each other will ensure higher possibilities of positive experiences at these attractions. By encouraging tourists to visit less popular destinations and sites, it may lead to those alternative attractions to become popular in their own right. This may ultimately lead to overtourism at the alternative attractions and so the cycle continues, reinforcing the need for sustainable attraction management initially and into the future.

Case study: Village Roadshow Theme Parks

The Gold Coast in south east Queensland, Australia is home to a range of theme parks and attractions, owned by two major theme park operators. Village Roadshow Theme Parks (VRTP) owns and operates Warner Bros. Movie World, Sea World, Sea World Resort, Wet'n'Wild, Paradise Country, Topgolf Gold Coast, Australian Outback Spectacular, and Village Roadshow Studios. Ardent Leisure operates Dreamworld and WhiteWater World (Scott, 2019). The theme parks are aimed at families, teens and young adults and provide 'world class shows, rides and attractions' (Gold Coast Bulletin, 2019). The CEO of VRTP, Clark Kirby said that the VRTP theme parks provide a 'truly encompassing experience' and that the company prides themselves on 'creating joy and memories that last lifetimes and we know that people are more likely to spend their money on shared-experiences where they can do just this' (Kirby, 2019).

In 2016 tragedy struck when four people were killed in an accident while on the River Rapid ride at the Ardent Leisure-operated Dreamworld. As a result, the Queensland government introduced new safety regulations which require a major inspection of rides by a qualified engineer every 10 years. Such a process will close each ride for several weeks to allow the thorough inspection of critical components to be undertaken. In addition, the operators are required to improve their record keeping for inspections, maintenance and operator competency and improve the competency and training of ride operators. In response to the new regulations Dreamworld introduced a new training academy, launched in March 2019. This training academy provides training programs for ride operators and loaders, which includes nationally accredited units of competency to ensure the employees have the skillset needed for these specific roles (Stone, 2019).

The tragedy at Dreamworld led to a decline in attendance at the Gold Coast theme parks by approximately 1.6 million people from the 2016 financial year to 2019 (Scott, 2019). The previous CEO of VRTP Graeme Bourke said of the tragedy that 'spooked mums and mums were nervous about putting their children on rides. It really hurt our Gold Coast business' (Australian Associated Press – Financial News, 2018).

To overcome this drop in visitor numbers, a $2.5 million domestic marketing campaign was launched in 2019 by Destination Gold Coast called 'Come and Play Gold Coast', with the hashtag #playgoldcoast which highlighted the many theme parks in the area as well as the dining, coffee culture and lifestyle (Mortimer, 2019). In addition, both of the Gold Coast's major theme park operators pledged to invest more funds into their assets to attract more visitors as they looked towards a 'recovery phase' (Scott, 2019). To this end, both operators developed multi-million-dollar masterplans to introduce new rides over the next few years (Pierce, 2019). Dreamworld's chief operating officer Greg Yong said:

> New technology in the theme park industry continues to develop and become more advanced each and every year. We believe the success of our future is about investing in new, state-of-the art experiences for our guests. This means some of our old favourites have been retired to make way for these new experiences (Pierce, 2019).

Both operators have invested in new rides, attractions and technologies to improve customer experience. Dreamworld is currently undergoing a $70 million redevelopment which includes introducing a new ride at a cost $20 million to develop. Called the Sky Voyager, the ride is described as the southern hemisphere's first flying theatre. In Sky Voyager the riders soar over an IMAX style screen and enjoy a five-minute flight of Australia's most spectacular scenery, complete with special sensory effects such as wind, sound, light, mist and scents (Stevens, 2019). VRTP has invested in a high tech guest experience management system which engages with guests at every part of their experience: from the moment they decide to purchase a ticket, during their time onsite, including virtual queuing, and once they have returned home as a way to provide 'real-time, personalized communication at every step along the guest journey' (PR Newswire, 2019). The CEO of VTRP said that these types of renewal projects and attractions 'demonstrate our dedication and commitment to the tourism industry and we anticipate visitors will come to the Gold Coast just for these experiences' (Gold Coast Bulletin, 2019). These developments reflect the need to continually upgrade rides to encourage repeat business and to ensure people believe they are receiving value for money, creating positive experiences and potential for excellent word of mouth recommendation.

Discussion questions

1 Thinking about a theme park that you have either visited or whose web page you have referred to, consider the means by which the attraction provides a memorable experience and what techniques were used to achieve this. In your answer you could consider the extent to which all five senses were used, the use of theming and the use of memorabilia.

2 The use of technology can enhance a visitor experience at a theme park. Discuss the extent to which technology has been used at a Village Roadshow theme park and the extent to which safety can be enhanced in the future by technology, as a means to encourage theme park attendees to return.

References

Australian Associated Press – Financial News (2018) Dreamworld on mums' minds: Village chief, *Australian Associated Press - Financial News,* 16 February, viewed 15/12/2019, www.sbs.com.au/news/dreamworld-on-mums-minds-village-chief

Bansal, A., Garg, C., Pakhare, A. and Gupta, S. (2018) Selfies: A boon or bane?, *Journal of Family Medicine and Primary Care,* **7** (4), 828-831.

Bilbao, R. (2019) Disney's new Star Wars: Rise of the Resistance to revolutionize ride technology, *Jacksonville Business Journal Online,* 3 December, viewed 13/12/2019, https://www.bizjournals.com/jacksonville/news/2019/12/03/disneys-new-star-wars-rise-of-the-resistance-to.html

Bleiberg, L. (2019) 10 awesome escape rooms in popular vacation spots, *USA Today Online,* 22 January, viewed 16/12/2019, https://www.usatoday.com/story/travel/destinations/10greatplaces/2018/09/14/escape-rooms-new-york-las-vegas-nashville-new-orleans/1256111002/

Business Wire. (2019) Global Amusement and Theme Park Market 2019-2023 - Rising Prominence of Virtual Reality (VR) Technology & Budding Augmented Reality (AR) Technology, ResearchAndMarkets.com, 13 December, viewed 15/12/2019, https://www.businesswire.com/news/home/20191213005106/en/Global-Amusement-Theme-Park-Market-2019-2023--

CE Noticias Financieras English (2019) Places that no longer want tourists, 17 September, viewed 14/12/2019 via Factiva.

Capocchi, A., Vallone, C., Amaduzzi, A., and Pierotti, M. (2019) Is 'overtourism' a new issue in tourism development or just a new term for an already known phenomenon? *Current Issues in Tourism,* 6 July, viewed 15/12/2019, https://doi.org/10.1080/13683500.2019.1638353

Carter, C. (2018) Bragging rights a travel goal, *Southland Times*, 4 September, viewed 12/12/2019, www.pressreader.com/new-zealand/the-southland-times/20180904

Cooke, E. (2019) Could mega theme parks help solve overtourism? *Telegraph Online*, 10 September, viewed 13/12/2019, https://www.telegraph.co.uk/travel/news/how-mega-theme-parks-could-solve-over-tourism/

Dann, G.M. (1981) Tourist motivation an appraisal. *Annals of Tourism Research*, **8** (2), 187-219.

Escape Mate. (2018) Wellington's very first fully mechanical escape room, *Scoop.co.nz*, 19 October, viewed 11/12/2019, http://business.scoop.co.nz/2018/10/19/wellingtons-very-first-fully-mechanical-escape-room/

Festinger, L. (1954) A theory of social comparison processes. *Human relations*, **7** (2), 117-140.

Financial Times (2019) Tourism: Instageddon impends, 1 September, viewed 20/12/2019, https://www.ft.com/content/de0fd360-c7ee-11e9-a1f4-3669401ba76f

Freitag, F. (2017). Like walking into a movie: intermedial relations between theme parks and movies. *Journal of Popular Culture*, **50** (4), 704-722.

Gold Coast Bulletin, (2019) Roadshow riding high, *Gold Coast Bulletin*, 14 November, viewed 11/12/2019, https://www.pressreader.com/australia/the-gold-coast-bulletin/20181114/textview

Groundwater, B. (2018) Treading lightly in overtrodden cities, *Sunday Age*, 8 July, pp 14-17.

Kirby, C. (2019) Q&A Clark Kirby, *Courier Mail, Queensland Business Monthly*, 27 September, viewed 11/12/2019, https://www.couriermail.com.au/business/as-the-man-anointed-to-take-over-village-roadshow-clark-kirby-has-to-balance-family-expectations-with-keeping-business-booming/news-story/c0ede13590098cd285f9814ef5f7ba9a

Kugel, S. (2016) The Getaway, Travel: How to get the most out of visiting world-famous sites, *New York Times Online*, 16 August, viewed 13/12/2019, https://www.nytimes.com/2016/08/21/travel/places-to-visit-attractions.html

MacDonald, B. (2019) Theme park of the future: Facial recognition replacing tickets at new Universal Studios in China, *Orange County Register*, 18 October, viewed 12/12/ 2019, https://www.ocregister.com/2019/10/18/theme-park-of-the-future-facial-recognition-replacing-tickets-at-new-universal-studios-in-china/

Macintyre, B. (2019) Tourist hotspots being destroyed by their success, *The Times*, 2 March, viewed 11/12/2019, https://www.thetimes.co.uk/article/tourist-hotspots-being-destroyed-by-their-success-t6zfvzv5s

Macau News Agency (2019) Diverting tourism fluxes to different attractions a good way for Macau to deal with overtourism, *Euromonitor*, 6 December, viewed 20/12/2019, https://www.macaubusiness.com/diverting-tourism-fluxes-to-different-attractions-a-good-way-for-macau-to-deal-with-overtourism/

Mahdawi, A. (2019) Meme tourism has turned the world into the seventh circle of selfie hell, *The Guardian*, 30 October, viewed 11/12/2019, https://www.theguardian.com/commentisfree/2019/oct/30/meme-tourism-has-turned-the-world-into-the-seventh-circle-of-selfie-hell

Matthews-King, A. (2018) No-selfie zones needed to stop people dying while trying to get the perfect shot, scientists claim, *Independent Online*, 30 September, viewed 10/12/2019, https://www.independent.co.uk/news/world/asia/selfie-deaths-falling-killed-pictures-drowning-driving-shoots-india-a8560836.html

Mortimer, L. (2019) Promote it and they will come, *Gold Coast Bulletin*, 23 April, viewed viewed 11/12/2019, https://www.goldcoastbulletin.com.au/business/gold-coast-theme-parks-experience-bumper-trade-during-easter/news-story/e00d070dc837a6f098c68034943f21ff

O'Connor, J. (2019) The Instagram decade: #hashtag holidays and other travel trends, *The Guardian*, 19 December, viewed 11/12/2019, www.theguardian.com/travel/2019/dec/19/instagram-decade-travel-trends-2010-2019-rail-glamping-airbnb-social-media

Parnell, K. (2017) That killer Instagram shot not worth dying for, *Daily Telegraph*, 22 October, viewed 12/12/2019, https://www.pressreader.com/australia/the-sunday-telegraph-sydney/20171022/textview

Pierce, J. (2019) Last hurrah for Tower of Terror, *Gold Coast Bulletin*, 24 October, viewed 10/12/2019, http://www.goldcoastbulletin.com.au/news/gold-coast/Flast-chance-to-ride-dreamworlds-tower-of-terror-before-gold-coast-theme-park-icon-closes/news-story/d03be8f12184363306941c9334302098

PR Newswire, (2019) Travel bloggers reveal 30 top emerging destinations to visit in 2020 if you want to help fix overtourism, *PR Newswire*, 5 November, viewed 13/12/2019, https://www.prnewswire.com/news-releases/travel-bloggers-reveal-30-top-emerging-destinations-to-visit-in-2020-if-you-want-to-help-fix-overtourism-300951447.html

Ross, S. (2019) Instagram-friendly maps launched for the Hebrides, *The Scotsman*, 10 July, viewed 10/12/2019, https://www.scotsman.com/heritage-and-retro/heritage/instagram-friendly-maps-launched-hebrides-1413550

Scott, J. (2016) Shocking amount of British tourists 'prefer taking selfies on holiday than actually exploring world famous sites' *Daily Mirror Online*, 4 June, viewed 10/12/2019, https://www.mirror.co.uk/news/uk-news/shocking-amount-british-tourists-prefer-8111277

Scott, M. (2019) Theme parks invest to lure visitors, *The Australian*, 5 September, viewed 13/12/2019, https://www.pressreader.com/australia/the-australian/20190905/282557314887720

Smith, S. P. (2019) Landscapes for 'likes': capitalizing on travel with Instagram, *Social Semiotics*, pp.1-29.

Stevens, K. (2019) Dreamworld's $70million makeover: Australian theme park vows to return to its former glory after River Rapids ride tragedy left it on the brink of collapse, *Daily Mail Australia*, 23 August, viewed 10/12/2019, https://www.dailymail.co.uk/news/article-7385717/Dreamworld-return-former-glory-River-Rapids-tragedy-left-brink-collapse.html

Stone, L. (2019) Queensland amusement parks ready for stricter safety regulations, *Brisbane Times*, 1 May, viewed 13/12/2019, https://www.brisbanetimes.com.au/national/queensland/queensland-amusement-parks-ready-for-stricter-safety-regulations-20190501-p51izr.html

Swan, M. (2019) The right social media platform for your library, in N.Verishagen, (ed.), *Social Media: The Academic Library Perspective,* Cambridge: Chandos Publishing, pp. 35-44.

Swarbrooke, J. (2012) *The Development and Management of Visitor Attractions* (2nd edn) Abingdon, Oxon: Routledge.

Tahseen, I. (2019) Obsessed with those travel selfies? You may be contributing to #Overtourism, *Times of India - Kochi Edition*, 21 November, viewed 12/12/2019, https://timesofindia.indiatimes.com/life-style/obsessed-with-those-travel-selfies-you-may-be-contributing-to-overtourism/articleshow/72071248.cms

Thurnell-Read, T. (2017) 'What's on your Bucket List?': Tourism, identity and imperative experiential discourse, *Annals of Tourism Research,* **67** (Nov), 58-66.

Torres, E.N., Milman, A. and Park, S. (2018) Delighted or outraged? Uncovering key drivers of exceedingly positive and negative theme park guest experiences, *Journal of Hospitality and Tourism Insights,* **1** (1), 65-85.

Travel Lemming (2019) 30 Top Emerging 2020 travel destinations, 5 November, viewed 13/12/2019 https://travellemming.com/2020-emerging-destination-awards/

UNESCO (2019) *World Heritage List,* viewed 22/12/2019, https://whc.unesco.org/en/list/

Weigold, M. (2018) Why do people risk their lives for the perfect selfie?, *The Conversation,* 13 May, viewed 10 December 2019, https://theconversation.com/why-do-people-risk-their-lives-for-the-perfect-selfie-55937

Wright, D.W.M. (2018) Terror park: A future theme park in 2100, *Futures,* **96**, (Feb), 1-22.

6 Events of the Future

Introduction

Events in their simplest form can be viewed as gatherings of people. Events have always existed as they allow people to gather for one or more of the following reasons:

- To solve problems individuals are not able to solve on their own;
- To celebrate;
- To mourn;
- To mark transitions;
- To make decisions because we need one another;
- To show strength;
- To honour and acknowledge;
- To build companies and schools and neighbourhoods;
- To welcome;
- To say goodbye (Parker, 2018).

Thus, a gathering can be described as the 'conscious bringing together of people for a reason' and it 'shapes the way we think, feel and make sense of our world' (Parker, 2018: i). This highlights the importance of events in society and confirms that the motivation to attend an event in the future is likely to remain the same because we are likely to continue to have the basic human need for inclusivity and contact (Hari et al., 2015). As human beings, the importance of social interaction is evident in our everyday life, we are shaped by other people and we crave social contact to the extent that 'isolation is used as punishment and even as torture' (Hari et al., 2015). Such face-to-face engagement may become more prevalent in the future because we spend 'more and more time in front of a screen each year' and so are spending less time engaging in face-to-face contact. As a result, 'face-to-face time has become a more treasured commodity in our modern world' (Social Tables, 2019).

The existence of online communities which are designed to facilitate face-to-face engagement between individuals is evidence of the importance of human contact. For example, 'Meetup' is described as an online platform for creating offline gatherings (Parker, 2018). The Meetup webpage explains that the organisation was created in 2002 as a platform to connect with other people in real life, and people use it to coordinate thousands of in-person meetings around the world for a range of purposes. The home page states 'The real world is calling. Attend local events to meet people, try something new, or do more of what you love' (Meetup, 2019).

Another human element that can come into play in relation to the future of events is the concept of FOMO or Fear of Missing Out. If event organisers recognise the existence of FOMO in an event setting, they should try to 'understand the experience from an attendee's perspective' (Alderton, 2019). To determine what kind of event would be satisfying and engaging for potential attendees, the event organiser should use social networks, conduct surveys and focus groups and listen to the event attendees prior to the event. By becoming 'audience-centric', the event organiser can develop stronger relationships with the attendee, build anticipation, give meaning to the event experience, and generate the feeling that if they don't attend, they will miss out on something special (Alderton, 2019).

Thus, the events sector is likely to continue to host gatherings of all sizes and types from birthday parties and weddings to business events, mega events and hallmark events. As well as the human desire for interaction and social engagement, the reason for the continued development of all types of events is that destination managers and governments recognise the importance of events as generating a range of socio-economic benefits (Dwyer *et al.*, 2016). However, event organisers need to reflect on the potential of their event to have negative impacts on the environment and plan accordingly, given the recognised sustainability issues associated with staging an event (Holmes *et al.*, 2015). In the future, mega events are likely to introduce new technology to run the event and to broadcast the event to the world. It may become even more competitive among potential host countries to host mega events in the future so the resources used in the bid process is likely to continue to grow (Greenwell *et al.*, 2019).

Event organisers should recognise that in the future when individuals travel to attend an event they are likely to continue to want to engage in sightseeing and experiences at the destination, to sample the food and learn about the culture and history of the area. Given that often the most memorable events celebrate local surroundings, it would be wise to expose the attendees to the local culture and connect them with the community to increase engagement

(BusinessMirror, 2019). This could be achieved by selecting an appropriate destination that provides high quality local dining experiences, cultural interactions and other activities which can be engaged in before or after the main event to be attended. For example, event organisers could incorporate city tours into the event itinerary or organise a scavenger hunt to encourage participants to explore the city (SpeedNetworking.com, 2018); and to organise authentic experiences in the local area in regards to art, food, attractions and the natural environment (Coppock, 2017).

Event sustainability and inclusivity

Events in the future will continue to grapple with issues concerning sustainability in such areas as the travel of attendees to and from the event (Jones, 2017), particularly since travel to attend events can generate significant greenhouse gas emissions. Technological advancements may enable conferences and business events to increasingly use distance attendance via video conference software. This may satisfy organisational and government policy on ensuring that staff reduce their carbon footprint. In addition, the event itself should be staged in an environmentally sensitive way (Getz, 2017). The greening of events has led to events introducing such measures as offsetting carbon emissions for event attendees, locally sourced food and drink, and the recycling of packaging and resources (Chirieleison *et al.*, 2019) with these aspects likely to continue to be of paramount importance in the future. Business event attendees will expect the event to be as paperless as possible, with minimal use of plastic, being mindful of food waste and donating what is left to local food banks. In addition to attempts to minimise energy use via turning off lights, switching off equipment, and not leaving projectors and computers running when they're not being used (SpeedNetworking.com, 2018). In addition, to keep the attendees healthy, break times can utilize the outdoors, or include body stretches, with low-carbohydrate options including snacks with less sugar to help attendees to focus throughout the day (Coppock, 2017).

Event attendees in the future are likely to continue to be socially and environmentally aware and so will be interested in learning what the event is doing in relation to these elements to ensure it has had minimal impact on the environment and has made a contribution to social good. For example, an event organiser could incorporate such initiatives as donating the floral centrepieces to a local children's hospital, use locally-sourced ingredients for food, or donating leftover food to food banks and shelters (Hall, 2016). Attendees could be offered healthy food, local and seasonal food to reduce food miles and perhaps individualised small plates to allow them to try a range of foods (Hall, 2016). In the future there is expected to be more of a desire to give back to the community with event activities and non-profit fundraisers (Coppock, 2017).

There is likely to be a growing concern among event managers in the future that some events are not inclusive in that some potential event attendees are not able to attend due to a range of constraints, such as those with family responsibilities or people with disabilities (Henderson *et al.*, 2018). Discussion is likely to continue in relation to how attendance at tourism and events could be viewed as a basic human right and should be open to all who want to attend (Veal, 2015). Thus, there is a growing movement towards ensuring that events such as business events and conferences are more inclusive to allow a greater diversity of people to attend these events. Such inclusivity can involve recognising the responsibilities of parents and accessibility for people with disabilities (Darcy, 2012). Diversity and inclusion relates to ethnicity and also age, gender, physical ability, religion, language, allergies and more. Events in the future will need to improve and enhance their ability to celebrate diversity and being inclusive (Colston, 2019). Since events should be inclusive, they must be planned to ensure they are accessible for everyone. This is reinforced by Alderton (2019) who said, 'all generations are part of our meetings now, and their needs and preferences overlap'. Event organisers should continue to recognise the importance of catering to attendees with food allergies and offer vegetarian, vegan, and nut-free, fructose free and gluten-free options (Coppock, 2017). There should be inclusive language on signage, better representation on panels, and clear codes of conduct about behaviour in regard to sexual harassment and bullying, as this would reflect an increased effort by the event organisers to create events that are more inclusive for everyone (Colston, 2019).

Business events in the future

Business events are generally understood to include meetings, incentives, conventions and exhibitions (the MICE sector) (Mair and Jago, 2010). Based on recent growth, business events are likely to continue to grow in the future (Pick, 2019), particularly because such events bring people together 'face-to-face in effective ways to help them achieve business and transfer ideas, to learn from one another and to build relationships' (Pelletier, 2017). There is also the development of the hybrid event whereby an event taking place at one location is able to feed into a much larger audience than a congress centre or hall can hold, reaching a much wider audience on a global level. Participants can engage in these events from their office, their home or when travelling on business. In addition, it allows highly regarded and renowned keynote speakers to be able to deliver their learnings from their home country (Arabian Business, 2015).

Understanding the drive for social interaction suggests that event organisers in the future should allow as much social interaction, networking and engagement at business events as possible. As mentioned above, people are looking for meaningful, personal ways to connect and share their story, as well as create direct business to business interactions (Coppock, 2017). In recognition of the need for face-to-face engagement and the seeking of networking opportunities, there is a need for event organisers to create opportunity for 'spontaneous conversations that come with serendipitous networking' (Social Tables, 2019). Ideas for networking can include speed networking sessions or round table discussions (SpeedNetworking.com, 2018). This serendipitous networking can be made easier by introducing software that helps match up people to networks who might not have otherwise met each other.

Event attendees in the future will continue to demand to be active event participants, with minimum lectures and passive observation, more discussion and active engagement (Pick, 2019). Sherrif Keramat, President and CEO of the Professional Conference Management Association (PCMA) said that people don't 'travel 2,000km to be in a square room in a hotel for three days and not experience anything. Young people especially want a sense of place. They want real, authentic experiences' (Ragavan, 2019). As a result, an event organiser has to think about how to create meaningful and memorable experiences for the attendees rather than just organising a series of educational presentations. Since the attendees want to be engaged, they want to learn hands on and want to interact with their peers so it's important to provide this for them (SocialNetworking, 2018). In addition, events can have more informal opportunities with attendees learning from each other, not just the instructor at the front of the room giving a lecture. This trend calls for more break time, break-out space and even a more casual rooming format which encourages informal interactions (Coppock, 2017). However, there is competition between similar events to ensure the event is as unique as possible to encourage more people to attend and to ensure the event is more memorable (Ralston *et al.*, 2007). Therefore, there is a need to research the needs and wants of event attendees to know how to design suitable events and as such, an increasing importance has been placed on event design (Orefice, 2018). Indeed, research into what makes an event successful and what makes an event a failure needs to be reviewed to allow event managers to make informed decisions about event design (Lade and Jackson, 2004; Nordvall and Heldt, 2017).

The choice of venue, content and activities are important at each event to ensure the event meets its goals (Pick, 2019). One suggestion for the future setup of a venue is to create 'ample meeting spaces' and 'collision spaces' (Social Tables, 2019). Having fewer chairs than there are attendees can encourage movement and interaction among event participants and help 'spontaneous interaction' (Social Tables, 2019) as lack of chairs allows attendees to freely mingle and not be isolated by sitting down or restricted to engaging with a limited number of people sitting within earshot. An event manager of the future may consider moving away from hotel venues to more purpose built venues such as convention centres or more unique venues to make the event more interesting, such as a studio or an industrial building, a farm or a museum, particularly if the location matches the event goals and event mood (Colston, 2019).

Use of event technology

The use of mobile applications (apps) at business and corporate events has grown exponentially (Pick, 2019), and this has been supported with technological developments from augmented reality (AR) and artificial intelligence (AI). In the future, mobile apps at events will continue to be used to plan an attendee's itinerary, to help with the check-in process and to answer questions, and will be able to collect valuable data about the attendee, which can be used by the event organisers (Pick, 2019). In addition, the mobile app can help personalise relevant content based on knowledge gained through their registration, onsite activities, previous engagements or their overall customer profile (Sherman, 2015). This allows event coordinators to connect with their audience in more meaningful ways and use data to provide personalised recommendations based on their interests (SpeedNetworking.com, 2018). Mobile apps can allow an event attendee to provide real-time feedback via live polling (Social Tables, 2019) and allow guests to give feedback in real-time. 'Attendees can let you know if the venue temperature is an issue, whether lighting is adequate for note-taking at a conference and if they are dissatisfied with the cleanliness of amenities' (Colston, 2019). By using data, the event organiser can make real-time changes, manage expectations and make adjustments, can immediately resolve a bad situation and subsequently save a negative experience from affecting many more attendees (Colston, 2019). In addition, before the event begins attendees can begin to engage in networking via internal social networks, such as Yammer, which can engage people before they arrive (Hall, 2016). Similarly, LinkedIn, Twitter and Facebook can allow attendees to share event content and discussions beyond the event.

As mentioned above, one of the top reasons participants invest in attending face-to-face events is the chance to network and learn from peers and industry experts. Until recently, it was difficult to facilitate networking at an event and draw true value from the activity. Most efforts left attendees wandering around events 'squinting at badges in a futile search to connect with the right person to advance their career or business' (Sherman, 2015). Today, new technologies within mobile apps can help enable networking and even make recommendations that facilitate one-on-one meetings (Sherman, 2015). In addition, with matchmaking technology, event planners can ensure their attendees are 'meeting with other like-minded people who share the same interests rather than hoping that it happens randomly' (SpeeedNetworking.com). Utilizing matchmaking technology makes networking more targeted and ensures better connections by attendees (SpeedNetworking.com, 2018). At the same time, it is wise to provide experiences that can create 'Instagrammable' social media moments (Colston, 2019), including photo booths and photo-sharing campaigns (SpeedNetworking.com, 2018).

Other software that can be used at events in the future to give immediate feedback are cameras and facial recognition software, which could 'measure sentiment (is everybody having fun?) in real time' (Pick, 2019). The software can obtain statements and opinions, attendees can vote and have their say on discussion topics and post comments and photographs as well as stream live videos. Other ideas of using technology to provide a greater event experience include the use of 'silence conferences' whereby multiple speakers can present in the same space and the attendees can wear headphones which allows them to toggle between speakers (Social Tables, 2019). The development of wearable technology including wristbands will provide the user with everything they need for their stay, such as their hotel key and managing their payment, reservation information and tickets, making the event experience seamless for the attendee (Colston, 2019). When networking, there is the opportunity to tap the wristband on name badges to facilitate an exchange of networking information. This will allow people to focus on the conversation instead of awkwardly exchanging hard copy cards and information (Colston, 2019).

In the future there is likely to be an increase in events which use a range of formats to allow event attendees to participate (Pearlman and Gates, 2010). Events such as conferences are likely to use a combination of formats such as hybrid, virtual and face-to-face (Hamm *et al.,* 2018). There needs to be recognition that some attendees may be more technology savvy than others and so a hybrid conference option may be preferable to cater to all event attendee needs. However, there is likely to be a growing awareness in the future of the importance of face-to-face engagement at events to ensure enjoyment (Hari *et al.,* 2015), recognising the power of meeting face-to-face and 'if you want to communicate and really connect with people, talking is a lot better than typing' (SpeedNetworking.com, 2018).

Safety and security

In the future at events, safety and security will continue to be important. Given that the event organiser often has access to the first and last name of attendees, titles, email addresses phone numbers, travel itinerary and hotel reservation, this reflects the importance of keeping information secure as it would be a gold mine for hackers (Colston, 2019). Event organisers will be expected to ensure protection of event attendee data, safeguarding guests both physically and online (Pick, 2019). Safety could be described as a basic expectation of event attendees where the 'best security is largely invisible' and protects the guests without impacting their experience (Pick 2019). In addition, event organisers must put in place crisis management communications, cyber security, and satisfactory onsite security to ensure the event can keep attendees safe. Event organisers should consider how best to communicate with attendees and stakeholders in the event of a crisis before, during or after an event as they develop their emergency and crisis communication plans.

The impact of Covid-19

There is no doubt that the international events sector has been severely disrupted by the Covid-19 pandemic, resulting in the cancellation or postponement of many events all over the world (Congrex Team, 2020). The most notable event postponed due to Covid-19 is the Japan 2020 Summer Olympics. The last time the Olympics were cancelled was in 1940 (Japan, rescheduled to Finland and then cancelled) and 1944 (London), during World War II (CBS News, 2020). Although global health concerns have existed in the past, in relation to the swine flu ahead of the 2010 Vancouver Winter Olympics, the bird flu before the 2018 Winter Olympics in South Korea and the zika virus during the 2016 Summer Olympics in Rio, none were considered serious enough to affect the scheduled timeline (NBC, 2020).

Event organisers need to balance consideration of both the health and well-being of attendees, staff and sponsors of these scheduled events along with the financial obligations of the event (Team Congrex, 2020). Event organisers can choose to cancel, postpone, relocate or where appropriate, offer a virtual substitute. While it is unlikely that face to face events will be completely replaced in the future with virtual events, especially within the meetings sector, it is expected that the current Covid-19 environment will 'accelerate a widespread switch to virtual and hybrid event formats' (Team Congrex, 2020). Preparing for a new landscape within the meetings and event sector in general, will require innovative strategies that consider how we connect with and attract event attendees as well as meet engagement and financial objectives (Team Congrex, 2020).

Summary

Events of the future will face similar issues in their design, planning and operation as they do at present because the inherent nature of the event will remain unchanged. Events will continue to be temporary in nature (Getz, 2008) and involve creativity, a range of stakeholders and extensive planning (Sautter and Leisen, 1999). However, the way that event planning and operational issues will be solved will reflect new technology and an enhanced understanding of the best way to resolve problems given that the area of event studies has matured (Lockstone-Binney and Ong, 2019).

As Millennials and Gen Zers continue to take up an increasing share of the workforce and as event attendees, there will be higher expectations for the use of technology at events to ensure the event is personalised, inclusive and well run, with excellent content and the opportunity to network with like-minded people (Pick, 2019). There should also be recognition that Millennial and Gen Z attendees are always on and always connected (Hall, 2019) so the use of social media and apps is likely to continue before, during and after the event. There should also be recognition that attendees at some business events may prefer some 'white space for mental regrouping and supporting participants' health and wellness' (Pelletier, 2017).

Case study: 2028 Los Angeles Olympics

In 2017 the International Olympic Committee (IOC) announced that the city of Paris would host the 2024 Summer Olympics, and that Los Angeles would host the 2028 Olympics. It is the first time the IOC has awarded a Games 11 years in advance, as it traditionally gives the next Olympic host city seven years to prepare. The announcement of both host cities at the same time was unprecedented, with the reason given that it would allow Los Angeles three extra years to plan for the event. Los Angeles has hosted the Summer Games twice previously; the first in 1932 during the Great Depression and the second in 1984, when it was the first modern games which did not leave behind a large public debt (People's Daily Online, 2017).

The IOC explained that Los Angeles was awarded the Olympic Games as they had articulated a vision to benefit Los Angeles citizens before and after the games. In the bid to be the host city, Los Angeles highlighted that there would be limited building work as many of the buildings that would be used already exist. After a three day visit to evaluate the Los Angeles bid, 'LA 2028', the IOCs commission was 'almost ecstatic' about the quality and standard of the venues they visited, and said of the venues, 'it goes from spectacular venues to impressive venues to mind-blowing venues to incredible venues'. One of the reviewers, Casey Wasserman, said 'It's certainly an incredible asset to be able to have a village that we could walk them through. It wasn't about greenfield sites or blueprints, it was about touching and feeling' (Axon, 2017). The Los Angeles Olympic plan is to use the city's prestigious UCLA campus as the Olympic and Paralympic Village, while the University of Southern California will house the main press centre and media village (Agence France Presse, 2019).

In response to a history of cost overruns that have left previous host cities with white elephants that cause budgets to skyrocket and serve little use beyond the Games, the IOC introduced the Olympic Agenda 2020 report. This is a series of reforms aimed at reducing the cost of bidding and making the Games more sustainable by encouraging the use of existing venues (Axon, 2017). The IOC said that they are impressed by the LA 2028 bid which demonstrated their sustainability priorities, reflected in their maximising the use of existing facilities (Daily Messenger, 2017). One commentator suggested that the Los Angeles bid can help to redefine the 'sustainability' of the Olympics, because the city does not need to build a single new permanent venue, neither in the Olympic village nor in the media village, as 'four sports parks around Los Angeles already have the existing venues for the games, which will translate into games with no incremental cost as well as a lasting definition of Olympics sustainability' (Daily Messenger, 2017). One bid official said, 'Our mantra was to fit the games to the city, not fit the city to the games' (Agence France Presse, 2019).

As part of that vision the 2020 Los Angeles Organisers indicated their plan to develop a youth sports programme in the decade leading up to the Games. In terms of legacy of LA 2028, the organisers' mission is to start an annual Legacy Triathlon event from 2019 onwards as a means to build the Games legacy before, rather than after, the event. By starting at the grassroots level, over time the event will be developed into a multi-day festival which will feature age-group and elite racing. The event will be designed to become a prestigious international event featuring elite and age-group athletes from around the world (Etchells, 2018).

On hearing the announcement, the Los Angeles Mayor Eric Garcetti said, 'This is a momentous day for the people of Los Angeles and a historic day for Los Angeles, for the United States and for the Olympic and Paralympic Movements around the world. For the first time in a generation, we are bringing the Games back to the City of Angels' (Business Day, 2017). He said 'Today we take a major step toward bringing the Games back to our city for the first time in a generation. LA 2028 will kick-start our drive to make LA the healthiest city in America, by making youth sports more affordable and accessible than ever before' (People's Daily Online, 2017).

The announcement of the successful bid in 2017 has allowed Los Angeles to have a 'decade of preparation', with the associated predicted economic influx from tourism, broadcast rights, licensing, inward investment, sales and oversees contracts and added post-event revenue. This is likely to 'catapult Los Angeles into a higher ranking of dominant world cities' with appreciation of commercial real estate and work/live urban neighbourhood revitalisation (Derhake, 2017).

The organisers announced in 2019 that the cost of hosting LA 2028 had risen to $US6.9 billion from the 2017 estimate of $US5.3 billion (Business Day, 2017). They suggested that the higher figure takes into account likely inflation over the next decade. The figure also factored in 3%, roughly $US200 million, the bulk of which will be invested in local youth sports and four additional years of operation (Agence France Presse, 2019). Revenues for the games include an $US898 million contribution from the IOC as well as $US637 million related to the IOCs sponsorship programmes. More than $US2.5 billion will be raised through domestic sponsorship in the United States along with a further $US1.93 billion in ticket sales and hospitality (Agence France Presse, 2019).

In 2018 the LA 2028 organisers began selling sponsorships, recognising that the event will need to compete for corporate money with the 2026 FIFA World Cup, celebrations around the 250th birthday of the United States and the emerging world of sports and gaming (Fischer and Lefton, 2018). NIKE is expected to be the first major sponsor of LA 2028 with a deal to include outfitting the American teams competing at LA 2028, as well as volunteer uniforms and LA 2028 merchandise to be sold to the public. Some industry experts have estimated that NIKE could pay as much as $US200 million to be a sponsor because of the merchandising, licensing and promotional potential of a deal with Los Angeles (Business Mirror, 2019). Another confirmed sponsor is Airbnb, who are keen to increase their sponsorship of major events as they see the benefits of local home owners providing accommodation for visiting event attendees (Masters, 2019).

Discussion questions

1 Thinking about the size and complex nature of preparing to stage an Olympic Games, what new or emerging technology could be utilised in LA 2028 to ensure the Games run smoothly and effectively?

2 Make suggestions on the means that the LA 2028 committee could use to ensure that before and after the games will provide maximum benefits to the people of Los Angeles.

3 Explain what the organisers of LA 2028 are planning to encourage inclusivity in the event.

References

Agence France Presse (2019) Los Angeles Olympics to cost $6.9 billion, *Agence France Presse,* 30 April, viewed 10/12/2019, https://sports.inquirer.net/347859/los-angeles-olympics-to-cost-6-9-billion

Alderton, M. (2019) BCD Meetings and Events Shares Hot Meeting Planning Strategies in New Report, *Meetings and Conventions*, 25 April, viewed 10/12/2019, http://www.meetings-conventions.com/News/Third-Party/BCD-Meetings-Events-Report-Meeting-Planning-Trends-Strategies-2019/

Arabian Business (2015) Events of the Future, *ArabianBusiness.com*, 16 March, viewed 13/12/2019, https://www.digitalstudiome.com/article-8726-events-of-the-future

Axon, R. (2017) Los Angeles 2024 Olympic venue plan impresses IOC committee in bid visit, *USA Today Online*, 14 May, viewed 13/12/2019, www.usatoday.com/story/sports/olympics/2017/05/14/los-angeles-2024-bid-impresses-ioc/101689700/

Business Day (2017) IOC crowns Paris 2024, Los Angeles 2028 in unique double, *Business Day*, 14 September, viewed 15/12/2019, https://businessday.ng/sports/article/ioc-crowns-paris-2024-los-angeles-2028-unique-double/

Business Mirror (2019) Nike set to become first major sponsor of Los Angeles 2028, *BusinessMirror*, 10 March, viewed 13/12/2019, https://businessmirror.com.ph/2019/03/10/nike-set-to-become-first-major-sponsor-of-los-angeles-2028/

CBS News. (2020) Japan's 2020 Summer Olympics – and billions of dollars – threatened by coronavirus pandemic, 12 March, viewed 7/8/2020, https://www.cbsnews.com/news/coronavirus-pandemic-2020-olympics-tokyo-cancelled-postponed-japan/

Chirieleison, C., Montrone, A. and Scrucca, L. (2019) Event sustainability and sustainable transportation: a positive reciprocal influence, *Journal of Sustainable Tourism*, **28** (2), 240-262.

Colston, K. (2019) 2020 event trends - how will the industry evolve?', *Endless Events*, 15 November, viewed 19/12/2019, https://helloendless.com/2020-event-trends/

Coppock, T. (2017) Meeting and event trends in a new year, *Herald-Times,* 6 June, viewed 14/12/2019, www.socialtables.com/blog/event-planning/event-trends/

Daily Messenger (2017) Los Angeles declares candidature for Olympic Games 2028, 2 August, viewed 16/12/2019, https://www.olympic.org/news/los-angeles-declares-candidature-for-olympic-games-2028-ioc-to-contribute-usd-1-8-billion-to-the-local-organising-committee

Darcy, S. (2012) Disability, access, and inclusion in the event industry: A call for inclusive event research, *Event Management*, **16** (3), 259-265.

Derhake, J. P. (2017) Benefits for Los Angeles what will hosting the 2028 Olympics Do for Los Angeles' reputation?, *National Real Estate Investor*, 14 August, viewed 13/12/2019, https://www.nreionline.com/finance-investment/what-will-hosting-2028-olympics-do-los-angeles-reputation

Dwyer, L., Jago, L. and Forsyth, P. (2016) Economic evaluation of special events: Reconciling economic impact and cost–benefit analysis, *Scandinavian Journal of Hospitality and Tourism*, **16** (2), 115-129.

Etchells, D. (2018) Proposed Los Angeles 2028 site to host annual Legacy Triathlon, *insidethegames.biz*, 27 May, viewed 13/12/2019, https://www.insidethegames.biz/articles/1065551/proposed-los-angeles-2028-site-to-host-annual-legacy-triathlon

Fischer, B. and Lefton, T. (2018) Nine years out, Los Angeles 2028 Olympic organizers to begin selling sponsorships, *L.A. Biz*, 31 October, viewed 16/12/2019, https://www.bizjournals.com/losangeles/news/2018/08/09/la28-marketing-arm-targets-2-5b-in-sponsorship.html

Getz, D. (2008) Event tourism: Definition, evolution, and research, *Tourism Management*, **29** (3), 403-428.

Getz, D. (2017) Developing a framework for sustainable event cities, *Event Management*, **21** (5), 575-591.

Greenwell, T.C., Danzey-Bussell, L.A. and Shonk, D.J, (2019) *Managing Sport Events*, 2nd edn, Champaign, IL: Human Kinetics.

Hall, A. (2016) 'New Ideas for Meetings in 2016: Flatbread, FOMO, Fresh Air', *MeetingsNet*, 29 January, viewed 16/12/2019, https://www.meetingsnet.com/financialinsurance-meetings/new-ideas-meetings-2016-flatbread-fomo-fresh-air

Hamm, S., Frew, E. and Lade, C. (2018) Hybrid and virtual conferencing modes versus traditional face-to-face conference delivery: a conference industry perspective, *Event Management*, **22** (5), 717-733.

Hari, R., Henriksson, L., Malinen, S. and Parkkonen, L. (2015) Centrality of social interaction in human brain function, *Neuron*, **88** (1), 181-193.

Henderson, E. F., Cao, X. and Mansuy, J. (2018) *In Two Places at Once: The Impact of Caring Responsibilities on Academics' Conference Participation: Final Project Report*. Coventry: Centre for Education Studies, University of Warwick.

Holmes, K., Hughes, M., Mair, J. and Carlsen, J. (2015) *Events and Sustainability*, Abingdon, Oxon: Routledge.

Jones, M.L. (2017) *Sustainable Event Management: A practical guide*, Abingdon, Oxon: Routledge.

Lade, C. and Jackson, J. (2004) Key success factors in regional festivals: Some Australian experiences, *Event Management*, **9** (1-2), 1–11.

Levinson, B. (2020) New survey reveals impact of COVID-19 on events industry, 2 June, viewed 8/8/2020, https://www.spicenews.com.au/industry-news/new-report-reveals-impact-of-covid-19-on-events-industry/

Lockstone-Binney, L. and Ong, F. (2019) Event studies: Progression and future in the field, in S. Beeton and A. Morrison (eds), *The Study of Food, Tourism, Hospitality and Events*, Singapore: Springer, pp.37-46.

Mair, J. and Jago, L (2010) The development of a conceptual model of greening in the business events tourism sector, *Journal of Sustainable tourism*, **18** (1), 77-94.

Masters, B. (2019) Airbnb is making a big bet with Olympic sponsorship', *Financial Times*, 19 November, viewed 11/12/2019, https://www.ft.com/content/73314fe2-0a01-11ea-b2d6-9bf4d1957a67

Meetup. (2019) viewed 23/12/2019, https://www.meetup.com/

NBC. (2020) Tokyo 2020 Olympics postponed over coronavirus concerns, 25 March, viewed 7/8/2020, https://www.nbcnews.com/news/world/tokyo-2020-olympics-postponed-over-coronavirus-concerns-n1165046

Orefice, C. (2018) Designing for events - a new perspective on event design, *International Journal of Event and Festival Management*, **9** (1), 20-33.

Parker, P. (2018) *The Art of Gathering: How we meet and why it matters*, New York: Riverhead Books.

Pearlman, D. M. and Gates, N.A. (2010) Hosting business meetings and special events in virtual worlds: a fad or the future?, *Journal of Convention & Event Tourism* **11** (4), 247-265.

Pelletier, S. (2017) 5 meeting trends that will prioritize people in 2018, *MeetingsNet*, 13 December, viewed 16/12/2019, https://www.meetingsnet.com/corporate-meetings-events/5-meeting-trends-will-prioritize-people-2018

People's Daily Online. (2017) Los Angeles declares candidature to host Olympic Games 2028, *People's Daily Online*, 1 August, viewed 15/12/2019, http://en.people.cn/n3/2017/0801/c90779-9249392.html

Pick, T. (2019) 10 Important Findings from the 2020 Future Event Trends Report, *Corporate Event News*, 1 October, viewed 19/12/2019, https://www.corporateeventnews.com/10-important-findings-2020-future-event-trends-report

Ragavan, S. (2019) PCMA highlights key trends for the year, *CEI Asia*, 3 May, viewed 10/12/2019, https://www.campaignasia.com/article/pcma-highlights-key-trends-for-the-year/450689

Ralston, L. S., Ellis, G. D., Compton, D. M. and Lee, J. (2007) Staging memorable events and festivals: An integrated model of service and experience factors, *International journal of event management research*, **3** (2), 24-38.

Sautter, E. T. and Leisen, B. (1999) Managing stakeholders: a tourism planning model, *Annals of Tourism Research*, **26** (2), 312-328.

Sherman, J. R. (2015) 3 corporate event trends for 2015, *Meetings & Conventions*, 24 March, viewed 13/12/2019, http://www.successfulmeetings.com/Strategy/Meeting-Strategies/3-Corporate-Event-Trends-for-2015/

Social Tables (2019) 19 Event Trends You Need to Know for 2020, *Social Tables*, 6 December, viewed 20/12/2019, https://www.socialtables.com/blog/event-planning/event-trends/

SpeedNetworking.com (2018) The Future of the Event Industry: 10 Things to Expect in 2019, *SpeedNetworking.com*, 11 November, viewed 10/12/2019, https://medium.com/@shannonkelly_80469/the-future-of-the-event-industry-10-things-to-expect-in-2019-1804fe169286

Team Congrex. (2020) Discruption in the business events industry: Rising to the challenges of COVID-19, viewed 7/8/2020, https://congrex.com/blog/disruption-business-events-industry-challenges-covid-19/

Veal, A.J. (2015) Human rights, leisure and leisure studies', *World Leisure Journal*, **57** (4), 249-272.

7 The Future of Tourism, Hospitality and Events Teaching and Training

Introduction

This chapter examines the ways in which teaching and training in tourism, hospitality and events have evolved and adapted to the contemporary demands of academia and industry. It explores the development of education in tourism, hospitality and events, the contemporary factors which influence teaching and learning, and discusses the rise of Massive Open Online Courses with a particular focus on their potential application within tourism, hospitality and events curriculum. The chapter concludes by providing an overview of Open Badges and their importance in education.

At the time of writing, the world has been confronted by the Covid-19 global pandemic which has caused great disruption at all levels. The impact of Covid-19 is briefly addressed in this chapter as the enforcement of social distancing measures has led to a significant increase globally in online education.

Education in tourism, hospitality and events

Tourism, hospitality and events (THE) are service-oriented sectors with a particular set of characteristics that determine the skills and competences required for graduates (Whitelaw *et al.*, 2009). These characteristics include inseparable production and consumption, the nature of guest relationships, labour intensive work, the cultural diversity of both staff and guests, and low barriers of entry to the industry (Whitelaw *et al.*, 2009). As such, skills generally associated with the workforce of these sectors include verbal and written communication skills, ethics, problem solving ability, leadership, critical thinking and strategic planning (Chapman and Lovell, 2006). At the educational level, curriculum design needs to be structured in a way that allows students to engage with multiple perspectives related to their subject matter, as well as gain the necessary graduate attributes to meet industry demands.

Initially, formal THE education focused on technical or vocational training (Inui *et al.*, 2006). Educational institutions provided training in relation to core competencies mainly related to hospitality and hotel management. Over time, the rapid growth in the demand for THE services generated the need for further diversification in levels of education. Institutions began to offer higher education degrees, including bachelor degrees with a management component and specialisations in particular areas such as Spa Management, Resort Management, and Sustainable Tourism. Higher education institutions have since moved from elite to mass providers of a diverse range of programmes in an attempt to align education with national policy relating to economic and social goals (Airey *et al.*, 2014). In many developed countries this has been achieved by providing accessible pathways for socio-economically disadvantaged groups or more economically viable education rates.

Tourism, hospitality and events undergraduate degrees often combine academic knowledge, vocational skills and competencies with some form of practical experience within the industry, usually through internships or Work Integrated Learning (WIL). However, studies have identified particular disparities between the objectives sought in academic curricula and those vocational skills required for work in the THE sector (Whitelaw *et al.*, 2009; Airey *et al.*, 2014; and Kim and Jeong, 2018). While similar issues exist for tourism, hospitality and events, these programmes have pursued different approaches (Airey *et al.*, 2014). Hospitality and event degrees have included vocational preparedness within the curriculum by providing training at restaurants or by coordinating events respectively. In the case of tourism, there is a greater focus on classroom activities with the inclusion of occasional field trips. The next section provides an analysis on the contemporary changes in THE higher education.

Contemporary teaching and learning in THE

Education in tourism, hospitality and events is currently facing significant challenges due to pressure from the global environment related to the changing needs of the industry, the dynamic nature of student markets and pressures on the academic workforce (Airey *et al.*, 2014). The internationalisation of education has also contributed to increased competition in the recruitment of international students as well as efforts to introduce national and international standards of qualifications.

The increased number of international students is generally perceived as a challenge for higher education institutions because these students have special needs regarding provision of language and social support. However, it should be noted that international students carry a set of experiences that potentially serve to enrich classroom discussions and activities. Kim and Jeong (2018) argue that the internationalisation of education could serve to foster cultural competency, as our diverse global society requires individuals who can positively interact with people from a broad range of social and demographic contexts (Chun and Evans, 2016). This is particularly important considering the diverse workforce and customer base that make up the tourism, hospitality and event sectors.

International programs include those provided by the Tuning Project and the International Centre of Excellence in Tourism and Hospitality Education (THE-ICE). The Tuning Project has led to the creation of the European Qualifications Framework which connects the qualification structures of participating countries. One outcome includes the standardisation of Bachelor, Master's and Doctoral programs that require institutions to identify the skills and knowledge each degree should deliver in their particular field of study (Dredge *et al.*, 2013). THE-ICE, on the other hand, validates programme accreditations and global quality assurance among a network of 37 higher education institutions (THE-ICE, 2020). At a national level in Australia, the Setting the Standards project has produced a set of nationally agreed upon and clearly articulated threshold learning outcomes (TLOs). These TLOs have been contextualized for THE courses in higher education, setting the minimum standards for any graduate in these programs and providing flexibility for institutions to include additional standards for their cohorts (Whitelaw *et al.*, 2015).

Considering the social impacts related to tourism, hospitality and events, Inui *et al.*, (2006) recommend a more comprehensive curricula inclusive of sociological topics concerning human interaction, social trends, sustainability and changing structures. They conclude this would engender more reflective students capable of critically examining the social responsibilities of these sectors in a range of areas such as poverty alleviation. Similarly, the service encounter component of tourism, hospitality and events would require students to develop attitudinal skills relevant to the context of their work as well as the organisation's strategic objectives and culture (Chapman and Lovell, 2006).

Increasingly, higher education institutions have begun to include experiential learning designed to provide students with a greater level of real-life business contexts (Kim and Jeong, 2018). Delivered as capstone projects, these subjects are effective in measuring knowledge and practical skills that can be transferred into the students' future professional paths. Other forms of experiential learning include case studies, community projects, field trips, and use of information technologies such as simulation software Hotel Management Simulations (HOTS). Other types of experiential learning include students participating in volunteering opportunities while studying in order gain exposure to practical experience and develop networks within the industry.

Technological factors also play a key role in adding an additional layer of complexity to the knowledge and skills required for graduates in these sectors. It is important that educational programs in these sectors address technological factors by preparing students with inter-personal (skills for interactions with others), intrapersonal (occurs in the mind of the individual), teamwork, as well as creativity and computer literacy skills (Whitelaw et al., 2009).

The rise of information and communication technologies (ICT) has changed the provision and delivery of services. While the use of ICT potentially adds value to an organisation, it also requires changes in institutional practices in order to adequately prepare students to become technologically competent. Cantoni et al., (2009) identified a range of providers that deliver degrees, courses or topics in tourism, hospitality and events using technological platforms. These included courses delivered in academic institutions; online training provided by a company or institution; destination management organisations (DMOs) that offer courses at local or national level; and, independent associations that provide topics aligned to each sector. The provision of these different educational avenues illustrates the impact of ICT on higher education within the THE sector.

The inclusion of advanced technology subjects in THE curricula is also discussed in Chapter 4 in relation to future workforce, robots in service delivery and their impact on the work environment, and appropriate management training (Murphy et al., 2017). The increasing adoption of robotics in tourism, hospitality and event operations is expected to disrupt the future of service delivery, resulting in the need to create new systems models (Bowen and Morosan, 2018; Martin, 2019). The demand for robots will address labour shortages and potentially contribute to generating effective and memorable experiences for customers. However, service employees will still be required as it is unlikely that robots will acquire social intelligence and communication skills to deal with complex emotional issues (Rosete et al., 2020).

Higher education in THE will continue to face challenges including increased competition for students among physical and online providers, changes to government higher education policy and reduced government financial support. As such, online curriculum design that allows students to engage in multiple methods of learning and applying knowledge will become vitally important. In early 2020, educational providers of all levels were faced with the immense challenge of conducting classes during a pandemic. Due to social distancing measures that countries implemented and fear of the spread of coronavirus, among other factors, universities and other educational providers transitioned their courses to online delivery within a very short time frame. This rapid transition may have profound significance for the future of education at all levels; however, emergent research will determine the extent of those potential changes. The next section will discuss the rise of online providers such as Massive Open Online Courses (MOOCs) and the potential of Open Badges in tourism, hospitality and events education.

The rise of Massive Open Online Courses

Massive Open Online Courses (MOOCs) which operate with a low or no cost access fee are available to anyone, and provide an 'affordable and flexible way to learn new skills, advance one's career and deliver quality educational experiences at scale' (Mooc.org, 2020). The earliest MOOCs were made available in 2008 (Sandeen, 2013) offering electronics and computer science courses targeted to people looking for high levels of professional development. Over time, universities in the USA began to offer MOOCs that were equivalent to the courses offered in the degree programs, although in some cases, students were denied academic credits upon completion as providers claimed to be protecting the integrity of the 'residential experience' (Sandeen, 2013). Today, MOOCs offered by some universities such as Helsinki University in Finland and Georgia State University in the USA offer students credit recognition. Platforms such as Coursera, Udacity or EdX, on the other hand, provide students with a form of authentication of completion (a certificate or badge) upon payment of a nominal fee.

MOOCs are perceived as revolutionary to education because their contribution to innovation in online pedagogy and data analysis has influenced the democratisation of education (O'Mahony and Salmon, 2014) and highlighted the potential for social inclusion due to their affordability (Sandeen, 2013). Some benefits provided by MOOCs include the development of independent and lifelong skills and access to specific programs and learning opportunities. MOOCs, however, are still considered in the realm of informal education and their accreditation or integration into conventional universities may be hampered due to perceptions that MOOCs lack the rigor in learning, assessment and identity verification of conventional onsite courses (Wang *et al.*, 2019).

MOOCs have evolved based on pedagogical models, namely cMOOCs and xMOOCs (Wang et al., 2019). The early versions, identified as cMOOCs, are driven by principles of social learning through the provision of free educational resources created by participants as primary content contributors. In contrast, xMOOCs, currently the prevalent category, hold a more unidirectional approach that relies on professors or teachers delivering information in a more traditional approach using technology available online. O'Mahony and Salmon (2014: 5) argue that xMOOCs are a type of 'disruptive technology', which refers to products or services that transform the traditional market and improve the product or service by offering cheaper prices or extending it to another category of consumer. To illustrate, xMOOCs (e.g., Coursera, Udacity or EdX) represent a form of market-oriented educational production that has shifted education to business entities and developed virtual brands which provide people with the opportunity to choose a course type or level and complete the course at their own pace and time.

In THE education, xMOOCs could play an important role in vital capacity building by developing higher order knowledge in collaboration with reflective industry practitioners (O'Mahony and Salmon, 2014). Universities could then work alongside xMOOCs to develop curriculum that scaffolds reflective skills such as critical thinking to allow graduates to engage effectively within their industry sectors. For instance, xMOOCs could provide further support in university-industry placements by providing a platform for integrative online assessments.

Digital or Open Badges

People are accustomed to having immediate access to information through a wide variety of online platforms such as the Internet and social media. Generating and sharing knowledge has provided opportunities for continuous learning within and outside educational institutions, and technology has changed the way we create, collaborate and innovate depending on our technological literacy. The way we learn has also expanded to include sophisticated 21st century skills that aim to ensure mobility between different working environments.

The increased demand for reskilling and aquiring lifelong learning capabilities is reflected in the expansion of online learning platforms utilised for training and education. This demand for gaining competencies and knowledge has resulted in the creation of digital credentials that capture learning through online tools. Digital or Open Badges have become an efficient means to evidence learning that occurs in customary ways but more importantly, provides recognition of those incremental skills not traditionally measured through formal assessments.

What are Open Badges?

Initially created by the Mozilla Foundation in 2010, Open Badges were designed to capture multi-dimensional learning and achievements through an online platform (Glover & Latif, 2013). Open Badges are akin to those worn by members of the international Scout Movement but in a digital format. The basis for the creation, award and display of these Open Badges emerged from a combination of digital games practices, online commerce systems (e.g., Amazon and eBay) and media culture (Gibson *et al.*, 2015). In effect, digital badges are integrated with technology through processes including learning analytics, real-time verification, customer management platforms, learning management systems and game mechanics (Grant, 2016).

Open Badges are visual representations with embedded metadata and images that validate a set of verifiable accomplishments. As such, they contain explicit data on what the individual learned, the institution responsible for awarding the badge, processes followed, date of award, validity period, activities undertaken, context and results achieved (Gibson *et al.*, 2015; Grant, 2016; Shields and Chugh, 2016). Open Badges can be used to visually display a skill, an educational qualification, a certification, community engagement, participation, authorisation and new skills not recognised by traditional educational providers (Open Badges, 2019). Learners are able to collect badges in a variety of ways including through online courses, gaming, sales and marketing initiatives, recognition from professional bodies and formal or informal assessments. An example of the implementation of these badges can be observed through MOOCs which offer this feature upon course completion. Once attained, learners are able to generate different sets based on their intended audience, such as a potential employer, supervisors, or other relevant stakeholders. The standards for the large range of badge-issuing platforms are established and managed by IMS Global, a not for profit global educational consortium.

Open Badges display distinct characteristics including an ongoing connection to sources for validation of data, evidence of achievements and an emerging set of standards outlining what constitutes a badge (Finkelstein *et al.*, 2013). These digital representations can then become portable and easy to display in different settings beyond paper certificates. In fact, due to its portability, digital badges can be stored and displayed in diverse online repositories, including e-portfolios and social media platforms such as Facebook or LinkedIn. The detailed information contained in an Open Badge provides the credential with transparency regarding abilities, skills and achievements. Organisations and communities that currently issue Open Badges include schools and universities, employers, community and non-profit organisations, government agencies, libraries and museums, event organisers, and companies and groups focused on professional development (Open Badges, 2019).

An important feature of these badges is their ability to display transferable and higher order skills not traditionally assessed or recognised, including problem-solving, digital literacy, team-work, critical thinking, creativity, entrepreneurial thinking and social skills. Open Badges can become an efficient tool which reflects academic achievements or prior learning as well as valuable competencies, therefore presenting an individual in a more holistic manner. Their representation of lifelong recognition also provides people from different ages with the opportunity to share a complete narrative of personal identity (Finkelstein *et al.*, 2013).

The increasing use of digital badging in promotion and engagement strategies provides users with status within the online community. This could potentially contribute to customer retention and loyalty (Gibson *et al.*, 2015). More importantly, the visual nature of badges allows the display of the incremental progress made by users and as such, turns milestones into observable achievements. This feature is particularly relevant for busy learners who may feel motivated to continue their learning pathway.

Despite Open Badges providing a potential solution for recognising academic knowledge and skills, some challenges have emerged, primarily in relation to the deployment of the badge and its properties (Myllymaki and Hakala, 2014). First, learners need to recognise the value of earning badges in order to serve as motivation for participation. Second, the content and criteria must be detailed and informative, particularly for those competencies that have limited validity. Last, the verification of a competency might require adjustments in teaching practices to allow teachers to verify whether a badge has been partially or wholly completed.

Open Badges in higher education

Used in a variety of fields, digital badges have become particularly relevant in education due to their potential to be designed for different pedagogies, curricula design and forms of assessments. Open Badges encourage learner engagement by allowing students to take control of their learning through offering flexibility in course selection, length of study and level of content. The transparency required to obtain a particular badge provides learners with autonomy through awareness of which activities to complete as well as opportunities to monitor their own progress. Open Badges provide users with incentives to engage in selected topics in greater breadth and depth further supporting learning opportunities (Glover and Latif, 2013; Shields and Chugh, 2016).

In terms of assessments, Open Badges can provide direct and indirect evidence of learning objectives while measuring competency and proficiency-based approaches. Badging can also allow institutions a more accurate mapping of credits from previous education; however, this must be supported by evidence of activities and experiences generated during the engagement (Gibson *et al.*, 2015). Evidence of activities and experiences could be provided through hyperlink to a description or a set of multimedia files encoded with metadata. While digital badges in education are rapidly evolving, their application in educational settings is a field yet to be methodically validated (Shields and Chugh, 2016).

The higher education sector in particular has begun to adapt to the demand for gaining competences and knowledge through employing open technologies. More importantly, the recognition that learners have acquired skills other than those displayed on degree certificates has contributed to the use of Open Badges as a complement to traditional degrees. The issue of employability has also contributed to the inclusion of advanced skills within the curricula as a way to confirm that individuals are prepared to perform a particular role. Employers now seek a combination of education and some form of tangible evidence that a person is equipped with the skills to perform a role. It is possible then to observe a strong focus on industry-based validation of curricula that explores new ways to develop human capital.

Institutions in higher education have implemented Open Badges initiatives in literacy classes, open online courses, MOOCs, co-curricular learning and local experimentation in classrooms (Grant, 2016). Gallagher (2019) noted that in higher education, Open Badges are generally included as part of capstone projects, work integrated learning (WIL) subjects or experiential learning opportunities. The inclusion of Open Badges in higher education seems a response to the potential shift from degree-based hiring to competency-based hiring. As a result, institutions have sought to include innovative models of learning and teaching to address this impending demand. Examples of inclusion of Open Badges at undergraduate and postgraduate level include:

- ☐ Illinois State University, USA, where Open Badges are integrated with a co-curricular transcript for a study abroad and service-learning program (Grant, 2016);
- ☐ University of Central Florida, USA, issues badges for information-literacy modules (Grant, 2016);

- ☐ The University of Notre Dame, Indiana, USA, integrates Open Badges into e-portfolios to recognise skills demonstrated by students through extracurricular and co-curricular activities (Diaz *et al.,* 2015);
- ☐ The University of California (USA), David's Agricultural Sustainability Institute's badge system based on proficiencies including competency and inquiry, strategic management, civic engagement and personal development;
- ☐ In Victoria, Australia, Deakin University through its subsidiary DeakinDigital offers a new model for workplace education and accreditation by providing workplace credentials, workplace learning and workplace qualifications (Deakinco, 2020); and,
- ☐ The Master of Science in Educational Design Technology at Concordia University, Wisconsin, USA, is a program based on more than 50 competency-based badges (Grant, 2016).

Different associations also provide support to higher education institutions by providing categories of digital badges that include community service, expertise development, presentation and facilitation, as well as leadership development (Educause, 2019). The Adobe Campus Leaders program issues educators with a range of badges related to their skills and abilities in digital media and creativity (Adobe Education Exchange, 2020).

In summary, Open or Digital Badges are a new way to capture and display information related to a person's knowledge and demonstrable skills and competencies. The potential for organisations to provide recognition beyond traditional transcripts is considerable, since Open Badges can serve to demonstrate a more detailed set of academic and personal accomplishments. In higher education in particular, badging can serve to unify formal assessments with informal and co-curricular activities. Open Badges could also have a significant influence on a learner's retention and overall satisfaction given they tend to foster motivation and engagement.

The future of global higher education and and training

At the time of writing the Covid-19 global pandemic has caused tremendous disruption to the international education system at all levels. Universities and other higher education providers have rapidly transitioned their courses to deliver content via online learning platforms, such as Zoom and Webex, in recent months.

Covid-19 has struck our education system like a lightning bolt and shaken it to its core. Just as the First Industrial Revolution forged today's system of education, we can expect a different kind of educational model to emerge from Covid-19 (Kandri, 2020).

The online learning landscape has been revolutionised and the desire from students for online delivery is likely to continue to grow as a result of Covid-19 (QS, 2020). Prior to the global pandemic, some higher education providers 'were seeing declines in enrolment for campus-based programmes and parallel increases in uptake of their online courses' (Kandri, 2020). Currently the 'online segment only comprises of a small fraction of the $2.2 trillion global higher education market' (HolonIQ, 2020), therefore great opportunity exists for higher education providers to transform their traditional delivery to incorporate online platforms. It is expected that some of the partnerships generated between universities, online education companies and tech providers will continue beyond the Covid-19 pandemic. Effective student engagement and adequate teacher training are challenges which will need to be overcome for future long term adoption.

Additionally, the importance of micro-credentials to address immediate upskill needs of the workforce during the pandemic has seen the introduction of shorter courses at a University level (Ross, 2020). In Australia, 64 short online courses provided by approximately 11 universities had been rolled out in alignment with industry needs by early May 2020 (Ross, 2020).

Summary

This chapter discussed the evolution of education in tourism, hospitality and events including two technological innovations – MOOCs and Open Badges – that present traditional educational providers with both challenges and opportunities. There is no doubt that the global Covid-19 pandemic has greatly impacted the higher education system and will spark a transformation of the sector to better meet the needs of providers themselves as well as domestic and international students. Further collaboration with current online providers will be necessary in order to present students with further opportunities for career development and the acquisition of relevant skills. Online education will be viewed as a complement, rather than a threat, to traditional higher education providers, and it is those providers who are best able to adapt to this changing environment that will reap the associated benefits.

Case study: Future industry training and development

Choosing to pursue a degree in THE requires the attainment of both theoretical knowledge and distinctive practical competencies. These sectors seek professionals with a strong focus on service provision who are at the same time knowledgeable about strategic management, finances, human resources and marketing. Future predictions indicate that these sectors will continue to grow (Grimaldi, 2019). Key issues set to accompany this growth include the notion of sustainability (see Chapter 10), the pervasive use of technology (See Chapter 4) and the focus on wellness and health tourism (See Chapter 9) to name a few. In addition, customers' needs will become more complex and diversified requiring different approaches to customer service as discussed in Chapter 4.

While pursuing their studies, future professionals are commonly presented with real-life scenarios and case studies that involve the identification of issues requiring potential solutions and recommendations. Through capstone projects or practice-oriented subjects, students acquire professional experience by working collaboratively with industry partners or community organisations. By working in teams, students are provided with opportunities to develop leadership and collaboration skills that are considered highly valuable when later applying for a job in the industry.

In many cases, students are commonly employed in the industry throughout the duration of their studies. This exposes them to a variety of opportunities to develop skills not necessarily reflected in their certificates of attainment. These soft skills may include time management, problem solving, emotional intelligence and cultural awareness. As such, THE students are able to gain substantial practical experience prior to the conclusion of their studies. The potential use of Open Badges represents an opportunity to capture those different skills and knowledge that students acquire throughout their educational experience.

Discussion questions

1. Considering the nature of THE industries, what do you think are the most relevant 21st century skills that will be required for future industry roles?
2. Digital badges are influenced by gamification elements, including competition with others and reputation within the group. How do you think this may apply to the tourism, hospitality and event sectors?
3. Badges should represent learning experiences that are relevant and motivating. Outline some of the professional competencies of the tourism, hospitality and event sectors that may be best represented in badges.

References

Adobe Education Exchange (2020), viewed 29/06/2020, https://edex.adobe.com/campus-leader

Airey, D., Dredge, D. and Gross, M. (2014) Tourism, hospitality and events education in an age of changes, in Dredge, D. Airey, D., and Gross, M (eds), *The Routledge Handbook of Tourism and Hospitality Education*, London: Routledge, pp. 3-14.

Bowen, J. and Morosan, C. (2018) Beware hospitality industry: the robots are coming, *Worldwide Hospitality and Tourism Themes*, **10** (6), 726-733.

Cantoni, L., Kalbaska, N. and Inversini, A. (2009) E-learning in tourism and hospitality: A map, *Journal of Hospitality, Leisure, Sport and Tourism*, **8** (2), 148-156.

Chapman, J.A. and Lovell, G. (2006) The competency model of hospitality service: why it doesn't deliver, *International Journal of Contemporary Hospitality*, **18** (1), 78-88.

Chun, E. and Evans, A. (2016) Rethinking cultural competence in Higher Education: An ecological framework for student development, *ASHE Higher Education Report*, **42** (4), 7-162.

DeakinCo. (2020) *About us*, viewed 4/05/2020, https://www.deakinco.com/about-us

Diaz, V., Finkelstein, J. and Manning, S. (2015) Developing a higher education badging initiative, *Educause Learning Initiative*, 5 August, viewed 8/03/2020, https://library.educause.edu/-/media/files/library/2015/8/elib1504-pdf.pdf

Dredge, D. Benckendorff, P., Day, M. Gross, M., Walo, M., Weeks, P. and Whitelaw, P. (2013) Drivers of change in tourism, hospitality and event management education: An Australian perspective, *Journal of Hospitality Education & Tourism Education*, **25** (2), 89-102.

Educause. (2019) *Microcredentialing*, viewed 8/12/2019, https://www.educause.edu/microcredentialing

Finkelstein, J. Knight, E., and Manning, S. (2013) The potential and value of digital badges for adult learners, Final Report, *American Institute for Research*, 16 July, viewed 8/03/2020, https://lincs.ed.gov/publications/pdf/AIR_Digital_Badge_Report_508.pdf

Gallagher, S. (2019) A new era of microcredentials and experiential learning, in *University World News*, 15 February, viewed 9/12/2019, https://www.universityworldnews.com/post.php?story=20190213103113978

Gibson, D., Ostashewski, N., Flintoff, K., Grant, S. and Knight, E. (2015) Digital badges in education, in *Education and Information Technologies*, **20** (2), 403-410.

Glover, I. and Latif, F. (2013) Investigating perceptions and potential of open badges in formal higher education, in Herrington, J., Couros, A. and Irvine, V. (eds), *Proceedings of World Conference on Educational Multimedia, Hypermedia and Telecommunications 2013*, Chesapeake, VA: AACE, pp.1398-1402.

Grant, S.L. (2016) Promising practices of open credentials: Five years of progress, *Mozilla*, viewed 2/03/2020, https://drive.google.com/file/d/0B7kHRuri9QdPQmRfdXZrblpSX0U/view

Grimaldi, E. (2019) Travel and Tourism continues strong growth above global GDP, *World Tourism Wire*, 27 February, viewed 9/12/2019, https://worldtourismwire.com/travel-tourism-continues-strong-growth-above-global-gdp-4038/#gsc.tab=0

Holon IQ. (2020) $74B online degree market in 2025, up from $36B in 2019, *Holon IQ*, n.d., viewed 15/05/2020, https://www.holoniq.com/notes/74b-online-degree-market-in-2025-up-from-36b-in-2019/

Inui, Y., Wheeler, D., and Lankford, S. (2006) Rethinking tourism education, in *Journal of Hospitality, Leisure, Sport and Tourism Education*, **5** (2), 25-35.

Kandri, S. (2020) How COVID-19 is driving a long-overdue revolution in education, *World Economic Forum – Agenda*, 12 May, viewed 16/05/2020, https://www.weforum.org/agenda/2020/05/how-covid-19-is-sparking-a-revolution-in-higher-education/

Kim, H.Y. and Jeong, M. (2018) Research on hospitality and tourism education: Now and future, *Tourism Management Perspectives*, **25** (Jan), 119-122.

Martin, F. (2019) What robots means for the future of the hospitality industry: A University of Houston Study examines the impact robots will have in hotels and restaurants, *Houston Public Media: Bauer Business Focus*, 8 March, viewed 3/04/2020, https://www.houstonpublicmedia.org/articles/shows/bauer-business-focus/2019/03/08/324596/what-robots-mean-for-the-future-of-the-hospitality-industry/

MOOC. (2020) viewed 3/04/2020, https://www.mooc.org/

Murphy, J., Hofacker, C. and Gretzel, U. (2017) Dawning of the age of robots in hospitality and tourism: Challenges for teaching and research, *European Journal of Tourism Research*, **11**, 104-111.

Myllymaki, M. and Hakala, I. (2014) Open badges in higher education, in Chova L., Martinez, A., and Torres, I. (eds) *EDULEARN14 Proceedings, 6th International Conference on Education and New Learning Technologies*, Valencia: IAETD Academy, pp 2027-2034.

O'Mahony, B. and Salmon, G. (2014) The role of Massive Open Online Courses (MOOCs) in the democratisation of tourism and hospitality education, in Dredge, D. Airey, D., and Gross, M. (eds), *The Routledge Handbook of Tourism and Hospitality Education*, London: Routledge, pp. 3-14.

Open Badges. (2019) *Issuing open badges*, n.d., viewed 9/12/2019, https://openbadges.org/get-started/issuing-badges/

QS (2020) How universities are embracing online learning during the coronavirus outbreak, *QS Quacquareli Symonds*, 26 March, viewed 15/05/2020, www.qs.com/how-universities-are-embracing-online-learning-during-the-coronavirus-outbreak/

Rosete, A., Soares, B, Salvadorinho, J., Reis, J. and Amorin, M. (2020) Service robots in the hospitality industry: An exploratory literature review, in Novoa H., Dragoicea

M., Kuhl N. (eds) *Exploring Service Science, IESS 2020, Porto, Portugal, February 5–7, 2020, Proceedings*, Cham: Springer, pp.174-186.

Ross, J. (2020) Australian universities, TAFEs dish up snack-sized courses, *Times Higher Education*, 1 May, viewed 4/05/2020, www.timeshighereducation.com/news/australian-universities-tafes-dish-snack-sized-courses#survey-answer

Sandeen, C. (2013) Integrating MOOCs into traditional higher education: The emerging "MOOC 3.0" era, *Change: The Magazine of Higher Learning*, **45**(6), 34-39.

Shields, R. and Chugh, R. (2016) Digital badges – rewards for learning?, *Education Information Technologies*, **22**, (Jul), 1817-1824.

THE-ICE. (2020) Our vision, mission, history and constitution, *The International Centre of Excellence in Tourism and Hospitality Education*, n.d., viewed 22/02/2020, http://the-ice.org/who-we-are/our-vision-mission-history/

Wang, X., Hall, A. and Wang, Q. (2019) Investigating the implementation of accredited massive online open courses (MOOCs) in higher education: The boon and the bane, *Australasian Journal of Educational Technology*, **35** (3), 1-14.

Wang, X., Hall, A. and Wang, Q. (2019) Investigating the implementation of accredited massive online open courses (MOOCs) in higher education: The boon and the bane, *Australasian Journal of Educational Technology*, **35** (3), 1-14.

Whitelaw, P.A., Barron, P., Buultjens, J, Cairncross, G. and Davidson, M. (2009). *Training needs of the hospitality industry.* Sustainable Tourism Pty

Whitelaw, P.A. Benckendorff, P. Gross, M.J, Mair, J. and Jose, P. (2015) *Tourism, Hospitality and Events learning and teaching academic standards*, Melbourne: Victoria University

Additional Resources

Coursera: https://www.coursera.org/

EdX: https://www.edx.org/

Massive Open Online Courses (MOOCs): https://www.mooc.org/

Open Badges: https://openbadges.org/get-started/

Setting the Standards Project: https://sites.google.com/site/tourismandhospitalitystandard/

Udacity: https://www.udacity.com/

Examples of game-based learning management systems:
Academy LMS: https://www.capterra.com/p/101970/Academy-Platform-LMS/
Axonify: https://www.capterra.com/p/130669/Axonify/

8 The Future of Film Tourism

Introduction

Film tourism refers to a post-modern experience at an attraction or destination which has been portrayed in 'some form of media representation, such as the cinema screen, television or video' (Kork, 2018: 5). Film-induced tourism occurs when a tourist visits 'a destination or attraction as a result of the destination being featured on television, video, DVD or the cinema screen' (Hudson and Ritchie, 2006: 256). Screen tourism, movie induced tourism and TV induced tourism are other terms commonly used in association with this type of special interest tourism (Riley *et al.*, 1998; Connell, 2005; Connell and Meyer, 2009). Beeton (2005), categorises film-induced tourism based on where the tourism activities occur, namely 'on-film' and 'off-film' induced tourism. 'On-film' induced tourism refers to tourism resulting from where a part of a film is shot and shown on the screen, while 'off-film' induced tourism refers to events or artificial destinations actualised through the involvement in films (Beeton, 2005). Film-induced tourism falls under the umbrella of cultural tourism as it represents the cultural heritage of a destination and may be considered an expression of visual arts and local traditions (Gjorgievski and Trpkova, 2012), with Kim *et al.* (2007: 1351) suggesting it has 'great potential to advance cultural exchange and understanding'.

As an illustration of the size and extent of movies which are filmed away from traditional studios, the 11 *Star War* movies filmed between 1977 and 2019 utilised a number of location sites throughout the world, in addition to sound film studios in the United Kingdom (Elstree and Pinewood Studios) and Australia (Fox Studio). Locations included the countries of Tunisia, United Arab Emirates, Norway, England, Guatemala, Switzerland, Australia, Italy,

Thailand, Spain, Ireland, Bolivia and the Maldives (Obias, 2018a). Only two USA locations were used for filming, both in California, however outside of the traditional Los Angeles and Hollywood sites. Similarly, many scenes in the six *Mission Impossible* movies spanning from 1996 to 2018 were filmed on location in various parts of the world, including Prague, Norway, Paris, London, Washington, Virginia, Sydney, Utah, Berlin, China, Vatican City, Morocco, Vienna, Kuala Lumpur, Budapest, Moscow Mumbai, Canada and Dubai (Looch, 2018; Obias 2018b). Table 8.1 shows a variety of on-film and off-film induced tourism.

Table 8.1: Examples of on and off-film induced tourism and runaway productions

Name of Movie or TV Series	Impacted Locations
Harry Potter	Kings Cross Station, London; Alnwick Castle, Alnwick, Northumberland; Glenfinnan Viaduct, Scotland; and, various locations in Oxford including Christ Church College, Bodleian Library and New College: United Kingdom
Lord of the Rings	Matamata, Waikato: New Zealand
Downton Abbey	Highclere Castle, Newbury, Hampshire: United Kingdom
Notting Hill	Notting Hill, London: United Kingdom
Braveheart	Stirling (despite being filmed in Ireland): Scotland
Four weddings and a Funeral	The Crown Hotel, Amersham, Buckinghamshire: United Kingdom (booked for four years following the movie release)
Field of Dreams	Baseball field, Dyersville, Dubuque County, Iowa: USA
Pretty Woman	Four Seasons Hotel, Beverly Hills Wilshire, California: USA

Source: Riley *et al.*, 1998; Goldstein, 2019.

Film tourism and destinations

Traditionally, Hollywood is the home of television and film making. However, in recent years there has been an increase in *runaway* films and television series, where filming occurs on location, usually at lesser known destinations (Johnson-Yale, 2008). Runaway productions are films and television series which are filmed in one country but initially released in another country. The interest among the public about the film helps to raise awareness of the site where it was filmed, which leads to the site becoming a potential attraction for domestic and international tourists. Franchise movies and television series, such as *Star Wars*, *Marvel*, *Game of Thrones* and *Mission Impossible* to name a few, have well planned filming and release date schedules which may enable destination locations to be scouted well in advance and subsequent tourism related opportunities be realised.

From a destination image and marketing perspective, a discrepancy may exist between the film-induced images and the images desired by the destination. For example, the way in which a destination is portrayed in a film may not be aligned with that of the destination management organisation (DMO). For example, the 1978 film *Midnight Express* (set in Turkey but filmed in Malta) and the 1982 television series *Bangkok Hilton* (set in Bangkok, Thailand) both portrayed their respective locations in a negative light. In addition, issues of inauthenticity and displacement within this context may arise when a movie is filmed in one place but is actually representing somewhere else entirely. Bolan *et al.* (2011: 106) considered a range of examples of displacement film tourism (or runaway production) including *Braveheart* (set in Scotland but filmed in Ireland), *The Last Samurai* (set in Japan but filmed in New Zealand), *Gangs of New York* (set in the USA but filmed in Italy), *Saving Private Ryan* (set in France but filmed in Ireland) and *Batman Begins* (set in the USA but filmed in England) just to mention a few. Although not all tourists expect a completely authentic destination experience, some do and may be disappointed when the reality does not match their expectations (Yeoman, 2008; Bolan *et al.*, 2011; Rickley and Vidon, 2018); White (2017) suggests that there are many film fans who would happily pay large amounts of money to visit an authentic location. Teng and Chen (2020: 7) suggest DMOs could consider strategies for increasing emotional attachment to the movie and the associated site. This could include activities such as providing exclusive fan clubs and celebrity activities as a method of 'reinforcing the fans' sense of belonging toward film celebrities'. White (2017) noted that it is really powerful if a destination highlights the 'emotional hook of the film's story' and connects that emotion to the location where their favourite movie was filmed.

Benefits of film tourism

The existence of film tourism suggests that by exposing the individual to the destination via the screen, they will be enticed to visit as a tourist (Saltik *et al.*, 2010). For example, one tourist from the USA was interviewed by Iannucci (2017) and said that she had a 'trip of a lifetime' when she visited Highclere Castle, West Berkshire where the television series *Downton Abbey* was filmed. She said 'I couldn't believe I was standing where the show was filmed.' Similarly, a couple from the USA were hooked on *Downton Abbey* as well as *Poldark*, and 'practically every British drama picked up by PBS'. The woman said, 'the pull of the gorgeously shot vistas (both internal and external) in these TV series was inescapable' (Iannucci, 2017).

Featuring a destination in a film or television series has the potential to increase visitation, generate visitor spending and increase the direct economic impact to the local economy. The increase in visitation creates the need for suitable destination development and planning in order to cope with the increased influx of film tourists and subsequently, minimise negative impacts (Croy, 2011). Local residents may also benefit from newly built tourism infrastructure and be provided with employment opportunities, ranging from catering to appearing in scenes as extras (Mordue, 1999, 2001). Long periods of location exposure, enhanced destination image creation through perfect scenes and actor appeal may also contribute to effective destination promotion and awareness. From a marketing perspective, the film or television series may potentially reach a greater audience than conventional advertising through their ability to appeal to a number of different market segments in a non-sales environment, as well as being a relatively easily accessed medium.

Destinations are continually seeking new ways to attract tourists in order to remain competitive. The use of a location at a destination for a movie or television series helps to reinforce the image of those places and has the potential to provide strong appeal. The film will ideally contribute to the overall destination perception and image development and subsequently, may be used to assist in forming a destination brand. A tourism authority could lobby film makers to encourage a sequel to a movie or future seasons of a TV show at their location, as both have the potential to allow the location to be popular among visitors by keeping the location fresh in their minds. For example, the historic mansion used in *Downton Abby*, namely Highclere Castle, Newbury received further attention when the TV series was made into a 'nostalgic, soapy movie'. The movie made more than $40 million at the box office but cost only $13 million to make so a future sequel is being discussed (Branco, 2019). The local businesses that have benefited from the increase in visitation in the vicinity of Highclere Castle include the local hotel, spa, golf course, restaurant and bar (Branco, 2019).

Private landowners whose sites have been used for film have also developed products to satisfy tourism demand. For example, *Harry Potter and the Philosopher's Stone* was filmed on location at Alnwick Castle in 2000. The following year, *Harry Potter and the Chamber of Secrets* (the second film in the series) again used Alnwick Castle for filming. In the movie it is on the grounds of the castle that Harry and his fellow students learned to fly broomsticks and learned the rules of the wizarding sport, Quidditch. In addition, the courtyards and baileys of Alnwick Castle were used to film Hogwarts students and staff going about their daily activities (Alnwick Castle, 2020). It is estimated that Alnwick Castle costs more than £1.5 million a year to maintain, so the owners have created various money-making ventures to cover their expenses. They open the house and gardens to visit during the summer months with a gift shop and tearoom, and offer 20 minute broomstick lessons included in the admission price so that 'young wizards can put their flying skills to the test in a fun-filled broomstick lesson for all, on the very spot where Harry had his first flying lesson in *Harry Potter and the Philosopher's Stone*' (Rowney, 2019). Over the years the owners have invited other film crews to film with the castle as a backdrop for *Blackadder, Robin Hood: Prince of Thieves, Elizabeth, Mary Queen of Scots, Transformers* and the Christmas special of *Downton Abbey*. But it is the Harry Potter experience that was the most popular. The 12th Duke of Northumberland and owner of the castle Ralph Percy said, 'There's been a huge Harry Potter effect and we're very grateful for it. We do as much Harry Potter stuff as possible and it just doesn't seem to die out' (Fryer, 2019).

Challenges of film tourism

While a number of destination benefits may be experienced from film induced tourism, potential challenges also exist. Increased visitation often brings increased negative social problems, including increased crime and disparities between the host population and visitors resulting in disgruntled locals beginning to detest the once welcomed tourists. For example, the popular film *Notting Hill* was released in 1999 and featured a small bookshop called The Travel Bookshop which was founded in 1979, but closed in 2011. Tourists are interested in visiting the Travel Bookshop as it was where the character of Hugh Grant in *Notting Hill* meets the character of Julia Roberts. However, since the Travel Bookshop used in the movie no longer exists, tourists now visit a similar small bookshop in the same area called the Book and Comic Exchange as they think it may be from the movie. They bombard the bookstore staff with questions about the movie and as a result the staff have been forced to post an angry sign to deter fans from asking about the movie. The notice reads, 'This is not the shop from the rubbish film Notting Hill. So don't ask! We don't know where it is either'! The confusion has become a source of irritation for the store's owners because the shop is small and can only hold a few customers at a time (Roundtree, 2018). In addition, the popularity of a destination due to film tourism may result in the physical carrying capacity of a destination being exceeded, ultimately causing damage to the local environment. For example, Maya Beach on the island of Koh Phi Phi Leh in Thailand, the location made famous by the 2000 film *The Beach*, has been closed between 2019 and 2021 in an attempt to allow it to recover from overtourism (Gunia, 2019). The location was previously receiving up to 5000 visitors per day and this resulted in damage to the coral reef and marine life in the local area.

The *Lord of the Rings* Hobbiton film site in New Zealand has proven wildly popular since its opening in 2002, and has put the town of Matamata firmly on the tourist trail and created jobs. However, its growth has put pressure on roads and other infrastructure as well as residents' goodwill, with one local complaining to the local council that tourists have been walking onto his property, taking photos of his house and cows, and using the roadside as a toilet stop (Tantau, 2019). The local council, the Matamata-Piako District Council, agreed that the Hobbiton Movie set would operate under their new development concept plan which involved a cap of 3,500 visitors per day for tours, working with the local council to monitor incoming traffic, visitor numbers and traffic management. This entailed installing extra road signage on local roads and investing in more sound systems to reduce noise (Hope 2019a). The local mayor of Matamata-Piako Jan Barnes said, 'We are very proud of Hobbiton, and we promote our district often through it. For the Hobbiton visitors it will be very exciting because often the tours were full. Now there will be more opportunities to experience the Hobbiton magic'.

The future of film tourism

In the future film tourists will be able to travel to be physically close to where the movie and TV action took place, and to stand in the physical space where actors stood because this gives them 'the chance to feel like part of the filming experience and to re-enact certain scenes' (Ghisoiu et al., 2017). Virtual reality enables travel to places that would otherwise be difficult to reach, but the question arises as to whether film tourists will be 'more motivated to be present physically rather than through a piece of hardware, regardless of how realistic it may render the space' (Bolan and Ghisoiu, 2020: 242). In the future an individual will have the option of experiencing film locations virtually in their own home environment rather than travelling to the original site. An individual's desire to engage with their favourite film or TV show could occur within an individual's home community when organisations develop movie or TV show related themed products and services for consumption in the home environment. For example, the development of a *Lord of the Rings* themed Airbnb in Ireland means that film enthusiasts do not need to travel to New Zealand (Middle Earth). Instead it allows visitors to stay in a Hobbit inspired room, located in a mountain location in Ireland, with copies of the Lord of the Rings DVD available to play (Delahaye, 2018).

More tourism related infrastructure will be required in the future to accommodate visiting tourists at various film tourism locations. For example, the Hobbiton Movie Set in the town of Matamata is New Zealand's third largest tourism destination. New direct scheduled flights from Auckland to Matamata have been scheduled to operate return journeys twice daily, seven days a week, with nine passengers per flight. The director of the airline Fly

My Sky, Keith McKenzie said, 'Often travellers in Auckland would like to go to places like the Hobbiton Movie Set, but don't have the time. This service will allow Hobbiton fans to visit the destination without the added hours on the motorway'. The flights from Auckland to Matamata for Hobbiton fans 'are expected to bring vibrancy and life' to a tired airfield (Hope, 2019b).

It is likely that in the future new niche film products will be developed for fans who want to visit film sites, and such products may be developed by local people who have insider information about the location of these sites. For example, in 2018 a company developed a movie location tour on Australia's Gold Coast as a three hour bus tour, visiting more than 20 film locations and 'celebrity hot spots'. In addition, the local tourism authority is planning to develop an augmented reality film location app that enables visitors to do self-guided drives or walking tours, with permanent markers at each location (Gold Coast Film Festival, 2019). Another example of the types of development which may increasingly appear in the future involves a dedicated film cruise which was developed in 2016 by Bolsover Cruise Club. This six month cruise was designed for film lovers and involved a round-the-world trip to visit 13 famous movies sites and involved travelling on 13 cruise liners with business class flight transfers. To ensure that all passengers were movie buffs, potential customers were required to prove themselves worthy of being part of the cruise by passing a 16 question movie knowledge entrance exam before being permitted to sign up. Michael Wilson, Managing Director at Bolsover Cruise Club, said, 'This cruise is for anyone who's dreamed of travelling the world and feeling like a film star at the same time. Filming locations often double as the most dramatic, scenic settings so we believe this voyage will prove to be a smash-hit' (Mcguire, 2016). Table 8.2 shows the locations travelled to during the film cruise.

Table 8.2: Locations visited during the film cruise with the associated movie

Location	Movie
Venice, Italy; and Barcelona, Spain	The Godfather
Rome, Italy; Athens, Greece; Istanbul, Turkey	The Good, The Bad and The Ugly; and, Mamma Mia
New York, USA	Spider Man
Los Angeles, USA	Pretty Woman
Hawaii, USA	Jurassic World
New Zealand	The Lord of the Rings
Phuket, Thailand	The Beach
Abu Dhabi, United Arab Emirates	Star Wars

Source: Mcguire, 2016.

Tourism destination authorities of the future would be wise to partner with location scout companies who specialise in finding appropriate locations for the setting of movies and TV series. For example, the company Easy Locations was established to help film producers and other productions in California find the perfect locations for their movies. The locations considered were homes, studios and commercial properties (restaurants, bars, warehouses, gas stations, stores, bakeries) which could be used as properties suitable for shooting movies, music videos, ads and documentaries. In addition, the home and business owners of these sites receive rent for the use of their property which could range from $2,000 to $10,000 per month depending on the size and quality of the property (Easy Locations, 2019).

In the future DMOs are likely to more strategically align their marketing strategies/image development with films and TV series and are likely to be better prepared from a destination and visitor management perspective. For example, Tourism Australia used the popularity of the 1986 block buster movie *Crocodile Dundee* to encourage US residents to visit Australia. In February 2018 a 'Son of Crocodile Dundee' marketing campaign was launched with an advertisement released during the Super Bowl match in February 2018 which was aired to an estimated audience of 110 million people (Tourism Australia, 2018). Two advertisements appeared during the Super Bowl match. The first advertisement was designed to look like an official trailer for an upcoming movie with the voice-over announcing that 'Dundee: The Son of an Australian Legend – Returns Home. This Summer. He'll embark on an Epic Adventure: In the Land Down Under'. The 60 second advertisement featured the Australian landscape, Qantas Airways and Akubra hats, and included appearances from nine well known Australian actors - namely Danny McBride, Chris Hemsworth, Hugh Jackman, Margot Robbie, Russell Crowe, Isla Fisher, Liam Hemsworth, Jessica Mauboy and Luke Bracey. The second advertisement repeated similar imagery to the first but included the original Crocodile Dundee (Paul Hogan) drinking at a Sydney bar, while it was revealed that the advertisement was not a trailer for a new movie but an advertisement for Australian tourism.

Immediately following the broadcast, Tourism Australia experienced record traffic to their web page. The two videos received more than 100 million video views on social media, and generated more than 12,000 media articles; the website was visited by people living in more than 10,000 US towns and cities, and delivered over $74 million in estimated advertising value (Tourism Australia, 2018). The effectiveness of the campaign was reflected by Qantas Vacations having increased website traffic and double the normal number of leads within the first 24 hours of the advertisement airing. Internet traffic to the web page of travel company Down Under Answers quadrupled the week after the launch, with ongoing enquiries being received (Tourism Australia, 2018). The campaign achieved a social media reach of approximately 890 million across multiple platforms, with 80% of this US-based. It was ranked in the top five best advertisements during the Super Bowl by Adweek, Bleecher Report and the Washington Post, and was the highest searched Super Bowl campaign this year according to Google (Tourism Australia, 2018). These figures reflect the success of the campaign and the impressive outcome of utilising the public's interest in the original movie to encourage further tourism.

Summary

As entertainment streaming services such as Netflix continue to produce popular programs and the movie industry continues to seek out suitable locations for filming, there may be a growth in film tourism to sites associated with movies and TV series. Although it is difficult to predict which movies, TV shows and associated sites will become popular, local tourism authorities should look to embrace film-related travel opportunities by anticipating what services and facilities to offer tourists that will provide a satisfying experience, particularly by tapping into the emotional aspect of the film or TV series. Timing is important to ensure that the greatest benefits are realised through increased visitation to the site while the movie or TV series is popular. The most popular sites to visit in the future will be ones that appeal to the visitor's emotions and these sites should relate to aspects of the characters and story lines. In a world of worries and anxiety, people like to be distracted from their problems through the escapism provided by film and TV series and this should be recognised and capitalised upon by tourism authorities to maximise tourism benefits.

Case study: The *Game of Thrones* tourist phenomenon

The first book in the series of fantasy novels *Game of Thrones* (GOT) is named *A Song of Ice and Fire*, and it was published in 1996 by American author George. R.R. Martin. Initially planned as a trilogy, Martin wrote and released a further four novels, with the sixth in the series due for release in 2020. The television series, consisting of 73 episodes broadcast over eight seasons, premiered in the USA in 2011. The popular series was ultimately broadcast in over 170 countries with more than 17 million viewers estimated to have tuned in to watch the premier of the eighth and final season. It is also estimated to have had a 'shadow audience' of another 54 million, reflecting those viewers who illegally downloaded it and/or viewed it later (Goldstein, 2019). Despite the series concluding in 2019, visitation to a number of the GOT filming locations, including Northern Ireland, Spain, Croatia and Iceland has continued to rise.

Northern Ireland is thought to have benefited most from the filming of a number of GOT scenes, contributing an estimated $US39 million annually to the nation's tourism industry (McDonnell, 2018). It was reported that the first four seasons of the television series generated a direct economic impact of approximately $US107 million to Northern Ireland's economy, in addition to the income earned from local employment and the actors' accommodation stays (Del Valle, 2019). Tourism Northern Ireland are confident that visitation in association with GOT will continue well beyond the airing of its final season and that further demand continues to exist for new visitor experiences (McDonnell, 2018). The Linen Mill Studios in Belfast already operates a GOT tour which includes visiting the Iron Islands and Giant's Causeway with a Rope Bridge Adventure, providing tourists with the opportunity to dress up in GOT costumes. The studio is set to open a new HBO 'Game of Thrones' studio experience in 2020. Currently, there appear to be no concerns regarding overtourism to these popular locations in Northern Ireland (Smith, 2019).

The small town of Gaztelugatxe (San Juan de Gaztelugatxe islet), located on the north central Spanish coast in the Basque country, is the location used to film the character Daenerys Targaryen's ancestral home of Dragonstone. The small town consisting of one main street and one café, has become inundated with GOT fans, and is often reported as being overrun with tourists in search of food and use of a toilet (Goldstein, 2019). Two weeks into filming some scenes for season five of the series in 2014, it was claimed by the US ambassador to Spain that local tourism had increased by 15% (Del Valle, 2019). Similarly, the Alcazar of Seville has grown in popularity due to featuring in the series as the capital of Dorne and its Water Gardens. (Smith, 2019).

In Croatia, a number of tours relating to the popular television series are advertised, including a walking tour of Dubrovnik which follows in the footsteps of the character Ayra Stark with a visit to Fort Lovrijenac (Kings Landing Dubrovnik, 2019). According to Dubrovnik's mayor, the show itself had accounted for half of the city's 10% increase in visitation in 2015 (Del Valle, 2019). The cruise industry is also contributing to increased tourist numbers with the city of Dubrovnik in 2017 becoming the world's second-leading cruise destination, receiving 750,000 visitors from 539 cruise ships (Del Valle, 2019; Goldstein, 2019). Dubrovnik began to experience a number of issues due to increased visitation and in 2017, introduced a 'Respect The City' plan. The main aim of this plan was to limit tourist numbers to the city's old town, classified as an UNESCO World Heritage Site, to 4,000 visitors at a time (Goldstein, 2019). Further comments made last year by the city mayor, Mato Frankovic, suggests that additional strategies may soon be introduced to slow down the number of tourists entering the city at the one time (Del Valle, 2019).

Iceland has been one of the fastest growing tourism destinations over the last nine years (Smith, 2019). The country received just 500,000 visitors in 2010 and this figure rose by 340%, to approximately 2.2 million in 2018, with the ratio of tourists to residents in Iceland now 7:1 (Smith, 2019). GOT has certainly had a major effect on tourism visitation to the country, along with the 2010 volcanic eruption of Eyjafjallajokull. In 2010 Iceland's economy was struggling due to the global financial crisis and now the popular town of Reykjavik is overrun with tourists. In Iceland several GOT locations, including Thingvellir National Park and Kirkjufell, better known as Arrowhead Mountain, are easily accessible and popular sites amongst tourists to visit, and visits to the sites of Vik and the Vatnajokull glacier featured during the final season are expected to increase.

The GOT themed tourism is expected to continue for some time with the many film locations remaining popular amongst 'set-jetters', and the recent announcement by the president of HBO programming that a prequel to GOT, *House of Dragon*, is currently being created (HBO, 2019). It is also likely that a number of these known film locations will be utilised as sets for other future movies and television series.

Discussion questions

1 Identify some advantages and disadvantages of 'runaway' productions, for both the film producers and the onsite locations.

2 How might destinations strategically use film in order to further develop their tourism visitation?

3 Assuming you are an individual who has never watched any GOT episodes, to what extent would you be interested in visiting these sites? Explain why you would or would not travel to these sites.

References

Alnwick Castle (2019) *Harry Potter*, viewed 21/01/2020, https://www.alnwickcastle.com/explore/on-screen/harry-potter

Beeton, S. (2005) *Film-Induced Tourism*, Clevedon: Channel View Publications.

Bolan, P. and Ghisoiu, M. (2020) Film tourism through the ages: From Lumière to Virtual Reality, in I. Yeoman and U. McMahon-Beattie (eds.), *The Future Past of Tourism: Historical Perspectives and Future Evolutions*, Clevedon: Channel View Publications, pp.236-252.

Bolan, P., Boy, S. and Bell, J. (2011) We've seen it in the movies, let's see if it's true: Authenticity and displacement in film-induced tourism, *Worldwide Hospitality and Tourism Themes*, **3** (2), 102-116.

Branco, N, (2019) The real Downton Abbey is magical, *Ottawa Sun*, 28 September, viewed 20/11/2019, https://ottawasun.com/travel/europe/the-real-downton-abbey-is-magical/wcm/0d6bd4f2-72f7-4974-a201-8f0163465839

Connell, J. (2005) What's the story in Balamory?: The impacts of a Children's TV programme on small tourism enterprises on the Isle of Mull, Scotland, *Journal of Sustainable Tourism*, **13**, 228-255.

Connell, J. and Meyer, D. (2009) Balamory revisited: An evaluation of the screen tourism destination-tourist nexus, *Tourism Management*, **30** (2), 194-207.

Croy, W. G. (2011) Film tourism: Sustained economic contributions to destinations, *Worldwide Hospitality and Tourism Themes*, **3** (2), 159-164.

Delahaye, J. (2019) You can now stay in a Hobbit-inspired Airbnb in Ireland from just £66 a night, *Mirror.co.uk*, 16 December, viewed 12/12/2019, https://www.mirror.co.uk/travel/uk-ireland/you-can-now-stay-hobbit-21106191

Del Valle, G. (2019) Game of Thrones tourism is wildly popular — and not just because the show is a hit, *Vox*, 4 April, viewed 23/12/2019, www.vox.com/the-goods/2019/4/4/18293965/game-of-thrones-tourism-croatia-iceland-ireland-spain

Easy Locations (2019) Scout and book the best movie locations: Search the largest database of movie locations in Southern California, viewed 19/12/2019, http://www.locations.film/

Fryer, J. (2019) What's it like to live at Hogwarts? *Mail Online*, 22 June, viewed 10/12/2019, https://www.dailymail.co.uk/news/article-7169061/Duke-transformed-glorious-Alnwick-Castle-gives-Mail-private-tour-blood-soaked-secrets.html

Ghisoiu, M., Bolan, P., Gilmore, A. and Carruthers, C. (2017) Conservation and co-creation through film tourism at heritage sites: An initial focus on Northern Ireland, *Revista Turismo & Desenvolvimento*, **1** (27-28), 2125-2135.

Gjorgievski, M. and Trpkova, S.M. (2012) Movie induced tourism: A new tourism phenomenon, *Journal of Economics*, **3** (1), 97–104.

Gold Coast Film Festival (2019) Gold Coast Movie Locations Tour: Light! Camera! Action, viewed 14/12/2019, https://www.gcfilmfestival.com/project/movie-locations-tour/

Goldstein, M. (2019) 'Game of Thrones' tourism growing even as the show ends, *Forbes*, 19 May, viewed 15/12/2019, www.forbes.com/sites/michaelgoldstein/2019/05/19/games-of-thrones-tourism-growing-even-as-show-ends/#2074ce5e1e29

Gunia, A. (2019) The Thai beach featured in the movie *The Beach* will be closed until 2021, *Time.com*, 9 May, viewed 30/12/2019, https://time.com/5587084/thailand-the-beach-closed-2021/

HBO. (2019) GoT Prequel 'House of the Dragon' gets the greenlight, *HBO*, n.d., viewed 23/12/2019, www.hbo.com/hbo-news/game-of-thrones-prequel-house-of-dragons

Hope, S. (2019a) Hobbiton expansion gets final tick to begin this December, *Stuff*, 4 October, viewed 9/12/2019, https://www.stuff.co.nz/travel/destinations/nz/116206298/hobbiton-expansion-gets-final-tick-to-begin-this-december

Hope, S. (2019b) Hobbiton flights benefit Matamata, *Waikato Times*, 27 November, viewed 5/12/2019, https://www.pressreader.com/new-zealand/waikato-times/20191127/281569472579014

Hudson, S. and Ritchie, J.R.B. (2006) Promoting destinations via film tourism: An empirical identification of supporting marketing initiatives, *Journal of Travel Research*, **44** (May), 387-96.

Iannucci, L. (2017) Take a vacation to your favorite movie locations; Here are some of the top spots, *MarketWatch*, 1 June, viewed 8/12/2019, www.marketwatch.com/story/take-a-vacation-to-your-favorite-movie-locations-2017-05-31

Johnson-Yale, C. (2008) So-called runaway film production: Countering Hollywood's outsourcing narrative in the Canadian press, *Critical Studies in Media Communication*, **25** (2), 113-134.

Kim, S.S., Agrusa, J., Lee, H. and Chon, K. (2007) Effects of Korean television dramas on the flow of Japanese tourists, *Tourism Management*, **28**, (5), 1340–1353.

Kings Landing Dubrovnik. (2019) viewed 22/12/2019, https://www.kingslandingdubrovnik.com/

Kork, Y. (2018) Films and tourist demand in C. Lundberg and V. Ziakas (eds), *The Routledge Handbook of Popular Culture and Tourism*. Abingdon, Oxon: Routledge.

Looch, C. (2018) The stunning locations of Mission Impossible', *Culture Trip*, 26 July, viewed 30/12/2019, https://theculturetrip.com/europe/united-kingdom/articles/the-stunning-locations-of-mission-impossible/

McDonnell, F. (2018) 'Game of Thrones' worth £30m a year to North's tourism sector, *Irish Times*, 10 May, viewed 23/12/2019, https://www.irishtimes.com/business/transport-and-tourism/game-of-thrones-worth-30m-a-year-to-north-s-tourism-sector-1.3490956

Mcguire, C. (2016) From Bond to The Beach: Film buffs invited on six-month cruise taking in some of the world's most famous movie locations, *Mail Online*, 14 July, viewed 22/12/2019, https://www.dailymail.co.uk/travel/travel_news/article-3688285/From-Bond-Beach-Film-buffs-invited-six-month-cruise-taking-world-s-famous-movie-locations-tickets-cost-104k-board-pass-TEST.html

Mordue, T. (1999) Heartbeat country: conflicting values, coinciding visions, *Environment and Planning*, **31** (4), 629–646.

Mordue, T. (2001) Performing and directing resident/tourist cultures in Heartbeat country, *Tourist Studies*, **1** (3), 233–252.

Obias, R. (2018a) 20 Star Wars movie locations you can visit in real life, *Mental Floss*, 5 January, viewed 30/12/2019, https://www.mentalfloss.com/article/65571/15-star-wars-movie-locations-you-can-actually-visit

Obias, R. (2018b) '20 Marvel movie locations you can visit in real life, *Mental Floss*, 23 April 2018, viewed 30/12/2019, https://www.mentalfloss.com/article/541194/20-marvel-cinematic-universe-movie-locations-you-can-visit-real-life

Rickley, J.M. and Vidon, E.S. (2018) *Authenticity & Tourism: Materialities, Perceptions, Experiences*, Bingley: Emerald Publishing.

Riley, R., Baker, D. and Van Doren, C. (1998) Movie-induced tourism, *Annals of Tourism Research*, **25**, 919-935.

Roundtree, C. (2018) No, we're not the bookshop from Notting Hill! , *Mail Online*, 9 May, viewed 12/12/2019 https://www.dailymail.co.uk/news/article-5708365/Bookshop-puts-angry-signs-tourists-think-store-featured-Notting-Hill-film.html

Rowney, J. (2019) Harry Potter filming locations you can visit in the UK, *mirror.co.uk*, 20 September, viewed 9/12/2019, https://www.mirror.co.uk/travel/uk-ireland/harry-potter-filming-locations-you-12990981

Saltik, I. A., Cosar, Y. and Kozak, M. (2011) Film-induced tourism: benefits and challenges for destination marketing, *European Journal of Tourism Research,* **4** (1), 44-54.

Smith, O. (2019) The Game of Thrones effect – how the TV series changed how we travel, *The Telegraph*, 23 March, viewed 23/12/2019, https://www.telegraph.co.uk/travel/destinations/europe/articles/game-of-thrones-effect-tourism/

Tantau, K. (2019) Enough of Hobbit hordes? Bad driving ? Pooing on the roadside? Selfies with the cows, *Waikato Times*, 9 April, viewed 10/12/2019, https://www.pressreader.com/new-zealand/waikato-times/20190409/281496457659672

Teng, H.Y. and Chen, C.Y., (2020) Enhancing celebrity fan-destination relationship in film-induced tourism: The effect of authenticity, *Tourism Management Perspectives*, **33**, (Jan) 1-11.

Tourism Australia (2018), Crocodile Dundee inspires new $36M American tourism push, *Tourism Australia*, 5 February, viewed 21/01/2020, https://www.tourism.australia.com/en/news-and-media/news-stories/crocodile-dundee-inspires-new-american-tourism-push.html

White, S. (2017) Tourism guru off to 'Star Wars' site, *Otago Daily Times*, 1 November, viewed 13/12/2019, https://www.odt.co.nz/news/dunedin/tourism-guru-star-wars-site

Yeoman, I. (2008) The authentic tourist, *Hospitality Net*, 4 August, viewed 15 /11/2019, https://www.hospitalitynet.org/opinion/4037066.html

9 The Future of Health and Wellness Tourism

Introduction

Wellness tourism is currently one of the fastest growing tourism niche markets having experienced exponential growth over the past two decades (Global Wellness Institute, 2018). The attributed reasons for the exponential growth is wellness being an essential factor in shaping people's lives, as well as being increasingly influential in patterns of consumption and production. The wellness industry plays a crucial role as an important driver for future business growth and major innovations (Voigt and Pforr, 2013; Pyke *et al.*, 2016). This chapter defines the relative terms of health, wellness, spa and medical tourism, identifies the current trends in the health and wellness sector, details the various wellness providers and considers the future direction of health and wellness in connection with tourism and destination development. The chapter concludes with a case study discussing the success factors of wellness spa tourism in Thailand.

Defining health, wellness, spa and medical tourism

Health tourism is a broad concept that comprises two major sub-categories of wellness and medical tourism (Mueller and Kaufmann, 2001; Smith and Puczko, 2009; Voigt, Brown and Howat, 2011). Smith and Puczko (2015) believe any definition of health tourism should ideally consider the World Tourism Organisation, which quoted in 1984 that health tourism was the extent to which 'an individual or a group is able to realise aspirations and satisfy needs, and to change or cope with the environment. Health is a resource for everyday life, not the objective of living; it is a positive concept, emphasising social and personal resources as well as physical capabilities'.

Kaspar (1996, cited in Mueller and Kaufmann, 2001: 7), who also regards wellness tourism as a subcategory of health tourism, defines the term as:

...the sum of all the relationships and phenomena resulting from a change of location and residence by people in order to promote, stabilise and, as appropriate, restore physical, mental and social wellbeing while using health services and for whom the place where they are staying is neither their principle nor permanent place of residence or work.

Smith and Puczko (2015: 206) believe the concept of wellness comprises life domains such as 'physical, mental and spiritual health, self-responsibility, social harmony, environmental sensitivity, intellectual development, and emotional well-being and occupational satisfaction'. These researchers (Smith and Puczko, 2015: 207) propose that wellness tourism extends beyond merely spa tourism (see definition below), with research indicating that wellness tourism also encompasses 'healthy cuisine, specific fitness or body-mind-spirit regimes, active ageing or longevity programmes, learning, adventure, spiritual enlightenment and personal growth, all of which take place in purpose-built centres'. Wellness tourism literature also discusses 'the absence of disease, illness and stress but also the inclusion of purpose of life, joyful relationships, and satisfying work and play through active involvement, presence of happiness, and a healthy body and living environment' (Smith and Puczko, 2015: 206).

The word 'spa' derives from the Latin phrase *sanus per aquam*, which means 'healthy through water' (Puczko and Bachvarov, 2006). A spa is a 'business offering water based treatments practiced by qualified personnel in a professional, relaxing and healing environment' (Lo, Wu and Tsai, 2015: 158). Though 'spa tourism' is the best-known form of health and wellness tourism, Smith and Puczko (2015: 208) argue as an actual tourism activity, that 'spa tourism' does not exist but rather tourists visit spas 'as places devoted to overall wellbeing through a variety of professional services that encourage the renewal of mind, body and spirit'. There are seven categories of spas including day spa, resort/hotel spa, destination spa, medical spa, mineral spring spa, club spa and cruise ship spa. In particular, the resort/hotel spa has the closest relationship to the tourism sector. Increasing spa visitation in recent years has increased both profitability and consumer awareness (Desseau and Brennan, 2008).

Connell (2006a: 1094) defines medical tourism as that which is 'deliberately linked to direct medical intervention and outcomes that are expected to be substantial and long term', inclusive of dentistry and plastic surgery. Connell (2006a) adds that the emergence of this tourism niche has helped to satisfy the needs of a growing number of people, mainly from developed countries, which has benefitted both themselves and various destinations, predominantly in developing countries. Laing and Weiler (2007: 381) consider a number of factors influencing the growth of medical tourism, including:

> ...the high cost of medical procedures, long waiting lists and ageing populations in wealthier countries, greater affordability of flights and travel, and a shift in medical care away from the public sector, such that people are more comfortable with paying for medical services offered by private bodies or companies.

The Internet, inclusive of online consumer forums and social media platforms, has also made retrieval of medical-related information regarding what is available, including comparison of prices and services, more straightforward. Being able to access more affordable medical facilities overseas gives individuals their anonymity, enabling surgery and recuperation to take place in a relatively unfamiliar environment (Connell, 2006b; Laing and Weiler, 2007).

The way wellness and medical tourism providers view health has consequences in regards to the products and services they offer and the kind of workforce employed (Voigt, 2013). Wellness and medical tourism cater to different tourists' needs and therefore different tourism markets. Each offer essentially differing types of services based in locations with very different characteristics and they have staff employing different skill sets (Voigt and Pforr, 2013). Weight loss and fitness camps for example, are aimed at overweight individuals and typically involve an educational programme integrating elements of health, nutrition and physical fitness, often within a specified timeframe (Gately *et al.*, 2000). Medical tourism providers are normally, but not always in the case of cosmetic and plastic surgery, illness-orientated and emphasise a curative approach. The majority of services offered are biomedical procedures including invasive and high-tech diagnostic services with personnel treating medical tourists (or patients), typically certified health professionals, such as doctors and nurses (Voigt, 2013).

Wellness tourism adopts an alternative approach to health that promotes, what Voigt and Pforr (2013: 3) believe is, 'the balance and holistic integration of multiple health dimensions inclusive of body/mind/spirit, environmental and social. Active self-responsibility, healthy lifestyles, subjectivity and actualisation of human potential also have a major role to play in wellness tourism.' Voigt (2013) observes that services offered by wellness tourism providers generally fall outside the realm of biomedicine, which can encompass a range of different services including Complementary and Alternative Medicine (CAM) therapies, which refers to the array of therapies that extend beyond Western medical treatments. The term complementary describes treatments used in conjunction with standard care, and the term alternative relates to less conventional methods of treatments (Good Therapy, 2020). Other services include non-invasive beauty treatments (cosmetic surgery), spiritual and lifestyle-based interventions. Additionally, staff caring for wellness tourists are normally not mainstream health professionals but instead consist 'of CAM practitioners, beauticians, nutritionists, and lifestyle coaches, or possibly religious personnel such as nuns and monks' (Voigt, 2013: 26).

This section has defined the key terms of health, wellness, spa and medical tourism. The remaining discussion focuses on the main historical trends in the development of wellness tourism, its core providers, and the megatrends considered to have driven the rising supply and demand of health and wellness tourism to date.

Development of wellness tourism

Wellness tourists typically are in search of some form of transformation (Voigt *et al.,* 2011) and on average will spend 53% more than the typical international tourist (Global Wellness Institute, 2018). Estimated at $639.4 billion in 2017, wellness tourism is a fast-growing tourism segment that has grown by 6.5% annually from 2015 – 2017 (more than twice the rate of general tourism). Tourists made 830 million trips in 2017, which is an increase of 139 million trips from 2015 (Global Wellness Institute, 2018). Wellness tourism has experienced several eras of growth and decline over the centuries. Voigt (2013: 23) argues that throughout history there has been 'an ever-changing interplay of two main functions: either the dominance of the pleasure dimension or the emphasis of a therapeutic principle.' Additionally, some researchers suggest that the dimension of spirituality also played a role in its development (Norman, 2011; Cusack, 2013; Voigt, 2013; Duntley, 2014; Csirmaz and Peto, 2015; Jepson, 2015; Norman and Pokorny, 2017).

From a Western perspective, Voigt (2013: 23-25) distinguishes four main trends (or eras) of wellness tourism, including:

1 Roman *balneums* (or community bathhouses) and the larger, more extravagant *thermae* which represent the first era. Romans perceived baths, particularly those fed by mineral or thermal springs, as medical centres where they sought cures for a wide range of diseases and aliments. In most cases, only Roman emperors and the very wealthy had the means and status to visit the thermae.

2 The second era of wellness tourism commenced during the Renaissance in Italy (15th and 16th centuries) where physicians, who (re-)discovered the value of water as a therapeutic modality in antique texts, started to promote balneology that involved the scientific study of the therapeutic

use of water in the form of bathing and drinking. While this triggered a revival in visitation to bathing establishments, the facilities were rather primitive and rustic compared to their Roman forerunners. To accommodate growing expectations by aristocratic and wealthy classes, who had the time and resources for travel and entertainment as never before, private and public investors in many destinations, particularly in England, Germany and the Hapsburg Empire, built glamourous spa and seaside resorts.

3 By the late 19th century, the expansion of spa and seaside resorts in European countries heralded the third era of wellness tourism. Despite spa development having declined in Great Britain, the expansion of entire townships and hospitality industries around European springs and along sea coastlines continued to accommodate steadily growing visitor numbers. Spa products and services became medicinal, scientifically analysed and researched in the curative branches of balneology, thalassotherapy (the study of therapeutic use of seawater) and fangotherapy (the study of the therapeutic use of mud and seaweed).

4 Due to recent sociocultural, economic and demographic change in the last two decades in particular, a fourth era of wellness tourism has begun spreading to most regions around the world, and in its wake, its products and its meanings, have diversified to include various wellness providers.

Figure 9.1: Roman Balneum in Bath, Somerset, UK. *Source:* C. Lade, 2020.

Providers of wellness tourism

Voigt (2013: 33 - 35) observes that while wellness tourism providers are not a homogenous group, there is nevertheless, three distinct core providers of wellness tourism:

1. Beauty spa hotels/resorts

The major focus of beauty spa hotels/resorts is the body and non-invasive beauty treatments, as well as a range of water-based and/or sweat-bathing facilities that sometimes utilise mineral or geothermal waters. Whilst there are exceptions, guests are usually passive recipients of the treatments provided to them. Many beauty spa hotels/resorts occupy the high end of wellness tourism offerings. Guests staying at five-star properties expect the provision of a spa on the premises. Day spas are similar to beauty spa hotels/resorts, the exception being they do not include, nor are they attached to, accommodation facilities.

Figure 9.1: Traditional Turkish Hammam in Antalya, Turkey. *Source:* C. Lade, 2020.

2. Lifestyle resorts

Also referred to as 'destination spas', the purpose of lifestyle resorts is to motivate guests to pursue a healthier lifestyle. These resorts cover a range of health-promoting domains such as nutrition, exercise and stress management. Guests commit to active participation as part of a very comprehensive program, usually tailored to their particular needs, identified at the start of their stay. Guests usually sign some form of contract where they commit to health-promoting behaviour and pledge to refrain from such activities such as smoking or drinking alcohol. The extensive schedule of activities, classes and workshops allows for relatively few free periods for leisure, with set mealtimes firmly based on a healthy diet. Amenities such as Internet access, televisions or telephones are not available, to ensure guests are not distracted. Many lifestyle resorts incorporate a beauty spa where treatment can be booked at an additional cost; however, the focus on beauty is not as pronounced as in the beauty spa category. Lifestyle resorts come in different sizes and levels of comfort; some being luxurious and opulent, whereas others have a more rustic barefoot luxury-feel to it, similar to that of 'boot camp'.

3. Spiritual retreats

The focus of spiritual retreats is on spiritual development. Spirituality is important, given its significance to many citizens, especially in Asia as well as those in Africa and South America. Smith and Puczko (2015: 206) believe spirituality, in the context of health tourism, best represents 'those forms of tourism that focus on physical health, but also improve mental and spiritual well-being and increase the capacity of individuals to satisfy their own needs and function better in their environment and society'. Spiritual retreats can be religious or non-religious but always include meditation in various forms. In most cases, specific teachings or philosophy and/or focus on the study of a specific mind/body technique such as yoga or Tai Chi form the basis of the meditation. Usually, a strict routine and time schedule governs the stay of guests. Compared to the previous two categories of wellness providers, spiritual retreats are more domestic where guests share rooms and/or bathroom facilities as well as having to participate in housekeeping routines such as washing dishes after meals and/or keeping their rooms clean and tidy.

Variations exist in relation to the supply and demand conditions of health and wellness tourism whereby influencing factors are inclined to differ depending upon nation (Csirmaz and Peto, 2015). In European nations, 'spas' are often considered in the context of medicinal purposes with government subsidy programs often provided while in North America, the term usually refers to beauty related recreation. Traditionally the United Kingdom, Germany, France, Spain and Italy were regarded as popular European health destinations however, in more recent years, countries such as Hungary, the Czech Republic, Romania, Poland, and Latvia have increased in popularity (Fink, 2018). Meanwhile, within Australia, North America and Canada, the focus is more inclined to be on promotion of a healthy lifestyle and prevention, with Australia and Canada in particular, emphasising their availability of natural resources (Csirmaz and Peto, 2015). Recreational day spas in the USA differ from those in Europe by incorporating a variety of treatments such as beauty, yoga, meditation, Tai Chi, massage, diet, fitness, intellectual and spiritual wellbeing. The Asian health tourist profile has changed substantially over the past two decades with a shift to that which is more representative of their USA and Australian counterparts. While meditation and spirituality, including the ancient art of Tai Chi, have always played a traditional role in health and wellbeing within Asian nations, there has been an increase in Asia in the USA trend for day spas located in shopping centres, beauty farms and fitness centres (Csirmaz and Peto, 2015). Before 1994 spas were not traditionally existent in South East Asia, however all major international hotels now offer their guests a spa experience. Voigt (2013) argues that the boundaries between the three types of wellness providers is not rigidly set. Instead, there is a gradual distinction between the differing types. For example, some lifestyle resorts focus more heavily on beauty treatment and massages, whereas others emphasise meditation and mind/body interventions or New Age services. Similarly, spiritual retreats may lean more towards lifestyle resorts by providing seminars addressing lifestyle issues, or by combining meditation with creative arts activities.

Megatrends in wellness tourism and future directions

Health and wellness tourism is fast growing with a focus on preserving people's health being a major influencing factor (Mair, 2005). Additionally, Csirmaz and Peto (2015) include demographic transformations, changes in women's social roles, enhancement of individualisation, spiritualisation and increased recognition and an appreciation for a healthy lifestyle as other major reasons for this increased growth.

Ongoing development of health related tourism over the last decade, especially dynamic development at a national and international level, has contributed to predictions of it becoming one of the fastest developing touristic and economic sectors (Ardell, 1985; Pilzer, 2007; Pyke *et al.*, 2016). In the past, destinations in Europe, North America and South East Asia recorded the most significant movement of tourists relating to health and wellness. Csirmaz and Peto (2015) add that their establishment was on a natural medical water foundation. However, 'product diversification, continuous innovation and the involvement of medical services into health tourism, the geographical embeddedness seems to disappear, thus basically restructuring the market' (Jonas-Berki *et al.*, 2014: 602).

Voigt and Pforr (2013: 4) define megatrends as those involving 'processes of transformation normally characterised by a half-life of at least 50 years that are apparent in all areas of life inclusive of everyday life, economy, politics and consumption.' These researchers (2013: 4-7) identify six megatrends they consider to have driven the rise of wellness tourism supply and demand, with the discussion framed in the context of future wellness tourism development:

1. Holistic health and increased health consciousness

Refers to individuals who are no longer passive recipients of health care. Holistic health considers how a person interacts with their environment. Rather than focusing on illness or specific parts of the body, this ancient approach emphasises the connection of mind, body and spirit (Guiding Wellness, 2020). Instead, people have become more informed and educated about their own health and the availability of contrasting health-care options. Developments in technology have made it much easier to access such health-related resources.

2. Pace of life acceleration

Where increased workloads and greater day-to-day pressures have seen a notable rise of stress-related disorders such as emotional strain (tension, fatigue, aggression) or psychological diseases (depression, burnout, anxiety). Wellness tourism represents an ideal outlet for people to escape from daily stressors and is a thoughtful way to relax and recuperate. The World Health Organization (2001) estimates that mental illness or a neurological disorder will affect one in four people worldwide at some time during their lives. Research indicates that factors such as stigma, discrimination and neglect, have contributed to preventing people from seeking available treatments with differences existing between Western and Eastern countries. (Cheng, 2015; Mirza et al., 2019; Krendl and Pescoszolido, 2020). Practicing mindfulness along with engagement in health-related activities is thought to serve as a preventative measure and contributes to maintaining mental health wellbeing overall. Development of mental health related organisations and/or campaigns such as 'Heads Together', championed by the Duke and Duchess of Cambridge, also paves the way in reducing the stigma traditionally associated with mental health wellbeing. High profile personalities and celebrities, such as Lady Gaga, openly discussing mental health wellbeing also assist in breaking down these preconceived notions surrounding mental health (Heads Together, 2019).

3. Inconspicuous consumption

Represents a counterbalance and a wide shift in consumer values away from fast-paced lives that are more materialistic, rationalised and technical. At this point, people decide voluntarily to simplify their lives by trading high incomes for more time and increased quality of life such as those seeking a 'sea change' in order to spend more time with family and focus on one's own wellbeing. Tourism bodies in Japan are developing new wellness trails in less populated travel routes also with a focus on 'forest therapy'. Forest therapy, translated as 'forest bathing' or *Shinrin-Yoku* in Japanese, concentrates on immersion of the body in the natural environment. This practice originated amongst the Japanese corporate community during the 1980s, in order to reduce stress levels and suicide rates. Gaining recognition worldwide, as much a preventative therapy as it is a treatment, countries including China, Korea, Taiwan, Germany, United Kingdom and the USA practice forest therapy (Forest Therapy Melbourne, 2019). The desire for healthier living, seeking a sense of meaning or joy and ways of stress reduction, in a way that is non-invasive and non-medical and is voluntary, all serve as potential motivations for wellness tourism.

4. Individualisation

With some trends suggesting people are seeking independent travel and niche tourism, individualisation in the context of wellness tourism has proved influential with the pursuit of self-actualisation, in accordance with Maslow's Hierarchy of Needs where an individual realises their full potential (Mcleod, 2020). Holistic wellness tourists seek to find their true selves, in the belief that the self (or its quest) is more important than external attractions and activities. 'Wellness tourist zones' may be the health trend of the future (Fratantoni, 2019). Destinations worldwide are creating these 'wellness tourist zones' primarily in order to reduce the effects of overtourism. Two well renowned health retreat destinations in Australia, Byron Bay in New South Wales and Daylesford in Victoria, offer tourists popular health related experiences in the form of spa and natural hot springs, yoga retreats and surf trips. Meanwhile in Croatia, there is development of a wellness and spa tourism zone in Varazdinske Toplice, a scenic region home also to the largest and oldest hot springs in Croatia (Croatia Traveller, 2019).

5. Spiritualty

Some consider the quest for spirituality to be the greatest megatrend in the 21st century. Despite the decline in religion, the yearning for spirituality and for meaning in life has not diminished. Although spiritual tourism is generally closely related to religious practices, Norman (2011) suggests that, rather, it may be considered in a socio-historical context, meaning a range of other 'non-religious' tourist activities may be incorporated. He specifically argues that the spiritual tourism desire for 'spiritual betterment' may be 'expressed as "problem solving" for everyday life concerns' (Norman, 2011: 203). For example, individuals may be walking the *Camino de Santiago de Compostela* in France and Spain to resolve a number of personal issues in their mind, and this now tourist activity impacts upon their happiness and well-being

(Norman and Pokorny, 2017). With spiritual concerns requiring attention during non-work time, it is only natural for the expectation for touristic breaks to be of spiritual orientations (Norman and Pokorny, 2017). The desire for such spirituality has seen a rise in the popularity of Eastern spiritual practices such as yoga and meditation, self-help books and pop psychology. Wellness and holistic health have seen a rise in the 'new age' movement.

6. Ageing population

Due to both longer life expectancies and falling birth rates, the world's population is ageing. The Baby Boomer (1946-1964) generation is currently driving this trend. Baby Boomers are generally more affluent, healthier and fitter than previous generations of similar ages and appear determined to stay in good health. Whilst this generation do not always constitute the largest number of all wellness tourists within all market sub-segments or in all countries, they are mainly responsible for increased demand for medical and wellness tourism providers (Voigt and Pforr, 2013). Ageing population is not only an issue for western society but also for eastern countries such as China. It is estimated that the number of people living in China over the age of 60 will be 248 million by 2020, with this figure increasing to 437 million by the year 2050 (United Nations, 2010). The current level of interest in health and wellness in China will subsequently only continue to grow, from its traditional focus on acupuncture, herbal medicine and hot springs, to encompass modern wellness programs and clubs (Heung and Kucukusta, 2013).

The impact of Covid-19 on health and wellness tourism

The Covid-19 restrictions enforced in many countries throughout the world have meant that many spas and health-related services have been severely disrupted. Regarded as non-essential services, many of these health-related businesses have been forced to close their doors, resulting in staff unemployment and reduced profitability.

Apart from the obvious physical health implications of contracting the Covid-19 virus and the reported potential long term health risks, the potential mental health implications of the Covid-19 restrictions themselves have been reported as significant. Factors including 'social isolation, anxiety, fear of contagion, uncertainty, chronic stress and economic difficulties may lead to the development or exacerbation of stress-related disorders and suicidality in vulnerable populations.......and people who have a family member or a friend who has died of Covid-19' (Sher, 2020:2).

The uncertainty associated with Covid-19, together with the different approaches implemented by governments across the world to try and contain the virus, have impacted to varying degrees the mental health and well-being of the global population (Health Management, 2020). Researchers at the University of Surrey, London School of Economics and Nottingham Trent University in the UK, led by Dr YingFei Hélio, are investigating the impact of these various government approaches on people's mental health and well-being. According to Dr Heliot, this study 'will aid prevention plans for policy makers across the globe in addressing mental health and well-being during and after the Covid-19 crisis' (Health Management, 2020:1).

Summary

Health and wellness encompasses a broad range of terms, categories, products and services. Shifts are occurring in the way in which people view their health and wellness, and importantly, how they incorporate wellbeing into their tourist related activities. While traditional health destinations remain popular, Eastern Europe, China and parts of South East Asia are beginning to capitalise on the health and wellness aspect and will only continue to expand to cater for the increased demand for related products and services. The world's ageing population will see an increased demand for wellness and medical tourism providers with attention directed towards stress reduction, mental health wellbeing and overall good health. Development of 'wellness tourist zones' is likely to be the way of the future, not only aimed at meeting the changed wellness needs of the tourist but also providing destinations with an alternate and sustainable means of development.

Case study: Wellness spa tourism in Thailand: Key elements of success

Wellness spas remain popular in Asian health-tourism destinations that focus on alternative treatments to achieve balance between mind/body/spirit. Treatments found in wellness spas include therapeutic massage, detoxifying clay baths, Ayurveda treatments (the ancient Hindu art of medicine and of prolonging life), colonic cleansings, acupressure, yoga, reflexology, meditation, aromatherapy, and hydrotherapy (Tourism Authority of Thailand, 2013). Laing and Weiler (2007) believe there are several reasons to explain why Asia is a global leader in health and wellness tourism. These include Western tourists seeking a return to nature and natural elements, the current trend towards traditional, more 'natural' treatments and therapies, and the affordability of cheaper international long-haul airfares relative to the disposable income of people in Western and other source markets. Thailand is one of the leading wellness spa destinations due mainly to the unique way in which the country is able to deliver a relaxing atmosphere, warm hospitality and service, and therapists with the 'soft touch' and welcoming personalities necessary for a rejuvenating, health-renewing wellness-spa experience (Tourism Authority of Thailand, 2013). Bangkok, Chiang Mai, Hua Hin, Koh Samui, and Phuket are home to many wellness spas catering to locals and travellers alike, as outlined by Thai media:

> One of the more famous spas in the Thai capital is Samsara Wellness, an oasis in the city providing a calm and relaxed atmosphere aimed at total wellness. In Chiang Mai, The Ayurvedic Centre has combined ancient Ayurvedic techniques with modern detoxification treatments in their pursuit of vitality for patients. Phuket is home to excellent wellness spas like Rarin Jinda, whose concept is to offer clients wellness, good health, and pampering in a sanctuary that is rooted in Thai massage and healing traditions. In addition, Sukko Cultural Spa and Wellness treats guests using ancient Thai medicine techniques such as traditional massages, heat therapies, and alternative treatments to bring one's body and spirit back to a state of harmony. Moreover, Aleenta Resort and Spas 'Mind Balance Wellness Retreat Package' rebalances the body's energy by removing toxins through three days of wellness activities and detoxification therapies
>
> (Tourism Authority of Thailand, 2013).

Han *et al.* (2018) believe that wellness spa operators in Thailand share five key success factors, including treatment quality, variety of service options, price, therapist and service staff, and facilities, each of which have helped build destination loyalty amongst its clientele.

1. *Treatment quality:* Measuring service quality consists of multiple dimensions, as explained in studies by Hansen (2014) and Lo *et al.* (2015). In the wellness spa business, the core products are the treatment programs (Han *et al.*, 2018). Studies of spa tourism, including Chi and Qu (2008) and Kucukusta and Guillet (2014), found the dominant quality of any wellness spa treatment is relaxation, which involves soothing the mind and refreshing the body, as well as relieving stress through wellness spa and massage experiences. Kiattipoom and Han (2017), in their examination of the destination attributes of Chiang Mai in northern Thailand, found that wellness spa tourists are generally seeking physical and emotional relaxation and stress management. Han *et al.* (2018) add that while wellness spa services differ from region to region, one constant feature of a spa across the globe is the physical and mental relaxation aspect.

2. *Variety of service options:* The availability of different service options is particularly important, as this is likely to determine international travellers' decisions and behaviours (Han *et al.*, 2018). Atienza *et al.* (2014), in their study of massage spa therapy in Batangas City in the Philippines, found that a variety of service options could comprise of various types of products, heterogeneity of treatments, and other medical and professional programs. Snoj and Mumel (2002) drew similar conclusions having

explored the importance of service quality in health spas in Slovenia. In Thailand, Han et al. (2018: 597) identified a diversity of wellness service programs including: spas (foot spas, soda fountains, hot tubs, aromatherapy, mud baths, peat pulp baths, steam baths, body wraps, and sauna baths) and massages/treatments include Thai herbal compress therapy, foot reflexology massages, and detoxification massages. Others include contour massages, abhyanga massages, skincare treatments, and anti-ageing treatments, which are essential constituents of various service options.

3 *Price:* Many international tourists travelling to Asia do so because of the reasonable prices for visiting a wellness spa, which Hashemi et al. (2015) observed in their study of spa and wellness tourism in Malaysia. Kucukusta and Guillet (2014) argue that prices can vary widely because of the diversity of products and services on offer at both local and international levels, though Kiattipoom and Han (2017) add that Thailand continues to attract an increasing number of overseas wellness tourists due to its price competitiveness.

4 *Therapist and service staff:* This aspect refers to a range of human-related factors such as responsiveness, empathy, interactional quality, proficiency, and skilfulness (Snoj and Mumel, 2002). Various studies including Bitner et al. (1994), Brady and Cronin (2001) and Parasuraman et al. (1988) have argued the importance of knowledge and courtesy of staff, and their readiness to provide a quality level of service. Han et al. (2018) observe that staff ability to perform the promised treatments and services in a dependable and accurate manner, as well as their willingness to provide genuine care and individual attention to guests are unarguably vital aspects of wellness spa tourism.

5 *Facilities:* These includes various tangible elements representative of the actual facilities including treatment room amenities, pools and jacuzzis, whirlpools, steam rooms, saunas, ambient physical environmental factors (serenity of atmosphere, temperature, music, and scents), crowdedness, design, decor, and furnishing (Kucukusta and Guillet, 2014). Han et al. (2018) also allude to the heterogeneity of facilities, equipment and surroundings according to the types of spa and massage products on offer.

Discussion questions

1 How important do you consider each of the success factors offered by wellness spa operators in Thailand? Classify each of the five factors from one to five, with one (1) being the most important, and five (5) being the least important. What are the reasons for your classification?

2 Undertaking an Internet search, what promotional information is available regarding Samsara Wellness in Bangkok, the Ayurvedic Centre in Chiang Mai, and the Rarin Jinda, Sukko Cultural Spa and Wellness, and Aleenta Resort and Spas in Phuket? What services do each of the centres offer? Does the promotional content entice you to want to visit one of Thailand's wellness spa destinations?

References

Ardell, D. (1985) The history and future of wellness, *Health Values*, **9** (6), 37–56.

Atienza, K. L. T., Evangelista, C. A., Kvangelista, L. I. and Ibre, R. T. (2014) Impacts of tourism industry of massage spa therapy in Batangas City, Philippines, *Asia Pacific Journal of Multidisciplinary Research*, **2** (5), 87–93.

Bitner, M. J., Booms, B. H. and Mohr, L. A. (1994) Critical service encounters: The employees' viewpoint, *Journal of Marketing*, **58** (4), 95–106.

Brady, M. and Cronin, J. (2001) Some new thoughts on conceptualizing perceived service quality: A hierarchical approach, *Journal of Marketing*, **65** (3), 34–49.

Cheng, Z. H. (2015) Asian Americans and European Americans' stigma levels in response to biological and social explanations of depression, *Social Psychiatry and Psychiatric Epidemiology*, **50** (5), 767–776.

Chi, C. G.-Q. and Qu, H. (2008) Examining the structural relationships of destination image, tourist satisfaction and destination loyalty: An integrated approach, *Tourism Management*, **29** (4), 624–636.

Connell, J. (2006a) Medical tourism: Sea, sun, sand and… surgery, *Tourism Management*, **27** (6), 1093–1100.

Connell, J. (2006b) Medical tourism: The newest of riches, *Tourism Recreation Research*, **31** (1), 99–102.

Croatia Traveller. (2019) Thermal Spa Resorts in Croatia, *Croatia Traveller*, n.d., viewed 22/12/2019, https://www.croatiatraveller.com/SpecialInterests/spas.html

Csirmaz, E. and Peto, K. (2015) International trends in recreational and wellness tourism, *Procedia Economics and Finance*, **32**, 755–762.

Cusack, C.M. (2013) History, authenticity, and tourism: Encountering the medieval while walking Saint Cuthbert's way, in A. Norman (ed.), *Journeys and Destinations: Studies in travel, identity, and meaning*, Newcastle-upon-Tyne: Cambridge Scholars Press, pp. 1–22.

Desseau, R. and Brennan, M.W. (2008) Spa feasibility: Steps and processes, in M. Cohen and G. Bodeker (eds), *Understanding the Global Spa Industry: Spa Management*, Oxford: Butterworth-Heinemann, pp. 110–129.

Duntley, M. (2014) Spiritual tourism and frontier esotericism at Mount Shasta, California, *International Journal for the Study of New Religions*, **5** (2), 123–150.

Fink, R. (2018) Health tourism in the EU: Facts and figures', *Regiondo Blog*, 4 April, viewed 7/09/2019, https://pro.regiondo.com/health-tourism-eu/

Forest Therapy Melbourne. (2019) viewed 19/07/2019, https://www.foresttherapymelbourne.com/

Fratantoni, M. (2019) Wellness tourist zones may be the health trend of the future, *New Daily*, 15 March, viewed 15/07/2019, https://thenewdaily.com.au/life/wellbeing/2019/03/15/wellness-tourism-trend/

Gately, P.J., Cooke, C.B., Butterly, R.J., Mackreth, P. and Carroll, S. (2000) The effects of a children's summer camp programme on weight loss, with a 10 month follow-up, *International Journal of Obesity*, **24** (11), 1444-1453.

Global Wellness Institute. (2018) *Global Wellness Tourism Economy 2018, Global Wellness Institute*, viewed 25/01/2020, https://globalwellnessinstitute.org/wp-content/uploads/2018/11/GWI_GlobalWellnessTourismEconomyReport.pdf

Good Therapy. (2020) Complementary and Alternative Medicine (CAM), viewed 14/07/2020, https://www.goodtherapy.org/learn-about-therapy/types/complementary-alternative-medicine?

Guiding Wellness. (2020) What is holistic health?, *Guiding Wellness*, n.d., viewed 1/05/2020, http://guidingwellness.com/wellness-and-holism/what-is-holistic-health?/

Han, H., Kiatkawsin K., Jung, H. and Kim, W. (2018) The role of wellness spa tourism performance in building destination loyalty: The case of Thailand, *Journal of Travel & Tourism Marketing*, **35** (5), 595– 610.

Hansen, K. V. (2014) Development of SERVQUAL and DINESERV for measuring meal experiences in eating establishments, *Scandinavian Journal of Hospitality and Tourism*, **14** (2), 116–134.

Hashemi, S. M., Jusoh, J., Kiumarsi, J. J. and Mohammadi, S. (2015) Influence factors of spa and wellness tourism on revisit intention: The mediating role of international tourist motivation and tourist satisfaction, *International Journal of Research – Granthaalayah*, **3** (7), 1–11.

Heads Together. (2019) viewed 10/11/2019, https://www.headstogether.org.uk/about/

Health Management (2020) COVID-19 and its impact on health and wellbeing, viewed 8/8/2020, https://healthmanagement.org/c/hospital/news/covid-19-and-its-impact-on-health-and-wellbeing

Heung, V.C.S. and Kucukusta, D. (2013) Wellness tourism in China: Resources, development and marketing, *International Journal of Tourism Research*, **15**(4), 346–359.

Jepson, D. (2015) The lure of the countryside: The spiritual dimension of rural spaces of leisure, in S. Gammon and S. Elkington (eds), *Landscapes of Leisure*, London: Palgrave Macmillan, pp. 202–219.

Jonas-Berki, M., Csapo, J., Palfi, A. and Aubert, A. (2014) A market and spatial perspective of health tourism destinations: The Hungarian experience, *International Journal of Tourism Research*, **17** (6), 602–612.

Kiattipoom, K. and Han, H. (2017) An alternative interpretation of attitude and extension of the value-attitude-behaviour hierarchy: The destination attributes of Chiang Mai, Thailand, *Asia Pacific Journal of Tourism Research*, **22** (5), 481–500.

Krendl, A. and Pescoszolido, B.A. (2020) Countries and cultural differences in the stigma of mental illness – The East West divide, *Journal of Cross-Cultural Psychology*, **51** (2), 149-167.

Kucukusta, D. and Guillet, B. D. (2014) Measuring spa-goers' preferences: A conjoint analysis approach, *International Journal of Hospitality Management*, **41** (Aug), 115–124.

Laing, J. and Weiler, B., (2007) Mind, body and spirit: Health and wellness tourismin Asia, in J. Cochrane (ed), *Asian Tourism: Growth and Change*, Oxford: Elsevier, pp. 379 – 389.

Lo, A., Wu, C. and Tsai, H. (2015) The impact of service quality on positive consumption emotions in resort and hotel spa experiences, *Journal of Hospitality Marketing & Management*, **24** (2), 155–179.

Mcleod, S. (2020) Maslow's Hierarchy of Needs, *Simply Psychology*, 20 March, viewed 10/05/2020, https://www.simplypsychology.org/maslow.html

Mair, H. (2005) Tourism, health and the pharmacy: Towards a critical understanding of health and wellness tourism, *Tourism*, **53** (4), 335–346.

Mirza, A., Birtel, M. D., Pyle, M. and Morrison, A. P. (2019) Cultural differences in psychosis: The role of causal beliefs and stigma in White British and South Asians, *Journal of Cross-Cultural Psychology*, **50** (3), 441–459.

Mueller H. and Kaufmann, E.L., (2001) Wellness tourism: Market analysis of a special health tourism segment and implications of the hotel industry, *Journal of Vacation Marketing*, **7** (1), 8–17.

Norman, A. (2011) *Spiritual Tourism: Travel and religious practice in western society*. London: Continuum.

Norman, A. and Pokorny, J.J. (2017) 'editation retreats: Spiritual tourism well-being interventions', *Tourism Management Perspectives*, **24** (Oct), 201–207.

Parasuraman, A., Zeithaml, V. A. and Berry, L. L. (1988) SERVQUAL: A multiple-item scale for measuring consumer perceptions of service quality, *Journal of Retailing*, **64** (1), 12–40.

Pilzer P. Z. (2007) *The New Wellness Revolution*. Hoboken, New Jersey: John Wiley & Sons.

Puczko, L. and Bachvarov, M. (2006) Spa, bath, thermae: What is behind the labels? *Tourism Recreation Research*, **31** (1), 83–29.

Pyke, S., Hartwell, H., Blake, A. and Hemingway, A. (2016) Exploring well-being as a tourism product resource, *Tourism Management*, **55** (Aug), 94–105.

Sher, L. (2020) The impact of the COVID-19 pandemic on suicide rates, *QJM: An International Journal of Medicine*, 1–6.

Smith, M. and Puczkó, L. (2009) *Health and Wellness Tourism*. London: Elsevier/Butterworth-Heinemann.

Smith, M. and Puczko, L. (2015) More than a special interest: defining and determining the demand for health tourism, *Tourism Recreation Research*, **40** (2), 205–219.

Snoj, B. and Mumel, D. (2002) The measurement of perceived differences in service quality – The case of health spas in Slovenia, *Journal of Vacation Marketing*, **8** (4), 362–379.

Tourism Authority of Thailand. (2013) Wellness Spa Tourism in Thailand, *ASEAN Tourism*, 11 October viewed 28/06/2020, https://www.aseantourism.travel/countryarticle/detail/wellness-spa-tourism-in-thailand

United Nations. (2010) *World Population Ageing 2009*, United Nations, 19 December, viewed 10/11/2019 , https://www.un.org/en/development/desa/publications/world-population-ageing-2009.html

Voigt, C. (2013) Towards a conceptualisation of wellness tourism, in C Voigt and C. Pforr (eds), *Wellness Tourism: A destination perspective*, London: Routledge, pp. 21–44.

Voigt, C., Brown, G. and Howat, G. (2011) Wellness tourists: in search of transformation, *Tourism Review*, **66** (1/2), 16-30.

Voigt, C. and Pforr, C. (2013) Wellness tourism from a destination perspective – Why now?, in C Voigt and C. Pforr (eds), *Wellness Tourism: A destination perspective*, London: Routledge, pp. 3–18.

World Health Organization. (2001) World Health Report: Mental disorders affect one in four people, viewed 8/8/2020, https://www.who.int/whr/2001/media_centre/press_release/en/#:~:text=Geneva%2C%204%20October%E2%80%94%20One%20in,ill%2Dhealth%20and%20disability%20worldwide.

10 Sustainable Development and Responsible Tourism

This chapter introduces the global sustainable development agenda and outlines the UN Sustainable Development Goals framework which is made up of 17 goals to be achieved by 2030. This global framework is offered as a lens through which sustainable development in the context of tourism is being implemented and measured. This is contrasted with responsible tourism, which is framed as the set of processes and mechanisms through which sustainable development is being pursued.

The case study of iSimangaliso Wetland Park in South Africa at the end of the chapter consolidates the key ideas from the chapter through highlighting the role that tourism plays in changing and shaping the socio-cultural and environmental fabric of societies, and sustainable approaches to addressing these changes today and in the future.

Introduction

The UN World Tourism Organisation (UNWTO) forecasts that the number of international tourist arrivals will increase from 1.5 billion in 2019 to 1.8 billion in 2030 (UNWTO, 2019; 2020e). Given these vast numbers, it becomes important to explore the existential threats to the tourism industry from a sustainable development perspective, and the impact this growth has on natural, socio-cultural and economic environments across the world.

It was Albert Einstein who said that we cannot solve tomorrow's problems with the same thinking we used when we created them. Given the complexity and 'wickedness' of the long list of today's sustainability challenges, which include: climate change, overtourism, land degradation, poverty, global pandemics, child labour and waste, responsible tourism requires a systems approach to tackle this complexity. This approach will be outlined later in this chapter.

The tourism industry is cognisant of these impacts and complexity, and sustainability in the industry is being driven by four forces of social change: dissatisfaction with existing products; growing environmental awareness and cultural sensitivity; realisation by destination regions of the precious resources they possess and their vulnerability; and the changing attitudes of developers and tour operators (Liu, 2003: 460). These drivers, individually and in unison, have created the impetus for the tourism sector to approach and address its stakeholders, including the natural environment, more sustainably.

The chapter begins by outlining the sustainable development agenda in the context of tourism and how responsible tourism practices are contributing and hindering progress on this agenda. Future approaches to dealing with unsustainable tourism are then outlined and conclusions drawn.

Tourism and Sustainable Development

The year 2017 was proclaimed as the United Nations (UN) Year of Sustainable Tourism for Development, in recognition of the tourism industry's capacity to contribute to sustainable development through its role in fighting poverty, promoting intercultural dialogue and conserving the environment (UNESCO, 2017). Sustainable development can be defined in numerous ways; however, one of the most enduring definitions is referred to as the 'Brundtland definition': 'Sustainable development is development that meets the needs of the present without compromising the ability of future generations to meet their own needs' (WCED, 1987: 41).

The Brundtland Commission was formed in 1983 in response to the significant deterioration of natural and human environments. *The Brundtland Report* (also known as *Our Common Future*) was produced at the culmination of the commission in 1987, and has popularised the definition of sustainable development presented above (WCED, 1987).

Since then, the global framework for international cooperation for sustainable development has been led by the 2030 UN Agenda for Sustainable Development and its 17 Sustainable Development Goals (SDGs). These goals are a call for action to end poverty, protect the planet, and ensure that all people enjoy peace and prosperity and overall transformative action towards sustainability. Attainment of the goals within the timeframe (by 2030) necessitates urgent, innovative, and far-reaching action from different actors – business, states, civil society, and individual citizens (UN, 2019).

The 17 goals are underpinned by 179 targets, and progress towards these targets will be tracked by 232 indicators (UN, 2019). Tourism can contribute indirectly or directly to all the SDGs, examples of which are presented below, along with the 17 SDGs.

Table 10.1: Examples of indirect and direct tourism contributions to the Sustainable Development Goals

1 NO POVERTY	Tourism can be linked to national poverty reduction strategies and entrepreneurship through low skills requirements and local recruitment.
2 ZERO HUNGER	Tourism can spur sustainable agriculture by promoting production, supplies to hotels, and sales of local products to tourists.
3 GOOD HEALTH AND WELL-BEING	Tax income generated from tourism and visitors fees collected in protected areas can be reinvested in health care and services.
4 QUALITY EDUCATION	Capacity and skills need to be built to ensure the tourism sector can prosper and provide job opportunities for youth, women and those with special needs.
5 GENDER EQUALITY	Tourism can empower women, particularly through the provision of direct jobs and income generation in tourism and hospitality-related enterprises.
6 CLEAN WATER AND SANITATION	Tourism investment requirement for providing utilities can play a critical role in achieving water access and security, hygiene and sanitation.
7 AFFORDABLE AND CLEAN ENERGY	Tourism can help reduce greenhouse gases (GHGs), mitigate climate change and contribute to energy access by promoting clean energy investments.
8 DECENT WORK AND ECONOMIC GROWTH	Decent work opportunities in tourism, particularly for youth and women, and policies that favour better diversification through tourism value chains can enhance tourism's positive socioeconomic impacts.
9 INDUSTRY, INNOVATION AND INFRASTRUCTURE	Tourism can influence public policies aimed at upgrading and retrofitting infrastructure to make it more sustainable, innovative and efficient.
10 REDUCED INEQUALITIES	Sustainable tourism can engage local populations and all stakeholders in tourism development and contribute to urban renewal and rural development.
11 SUSTAINABLE CITIES AND COMMUNITIES	Tourism can, inter alia, promote urban regeneration, and preserve cultural and natural heritage.
12 RESPONSIBLE CONSUMPTION AND PRODUCTION	Adopting sustainable consumption and production (SCP) models can help monitor sustainable development impacts for tourism, including for energy, water, waste, biodiversity and job creation.

13 CLIMATE ACTION	Tourism stakeholders can play a critical leading role in fighting climate change by reducing their carbon footprints.
14 LIFE BELOW WATER	Tourism development can help preserve marine ecosystems and promote a blue economy and the sustainable use of marine resources.
15 LIFE ON LAND	Sustainable tourism can help conserve and preserve biodiversity, and generate revenue as an alternative livelihood for local communities;
16 PEACE, JUSTICE AND STRONG INSTITUTIONS	Tourism can help foster multicultural and interfaith tolerance and understanding, and peace in post-conflict societies.
17 PARTNERSHIPS FOR THE GOALS	Tourism can strengthen public-private partnerships (PPPs) and engage all stakeholders to work together to achieve the SDGs.

Source: UNWTO and UNDP, 2017: 16-17.

Sustainability is often associated with the preservation and rehabilitation of the natural environment or being 'eco-friendly'. However, the SDGs and the tourism sector's contribution (See Table 10.1) highlight that the interconnected and mutually reinforcing three pillars of sustainability – the socio-cultural, environmental and economic pillars, are more useful for understanding the complete picture. For example, environmental challenges such as climate change and deforestation have implications for society and economies across the world through their effects on communities and their livelihoods respectively. Rural farming communities' livelihoods in numerous parts of the world are affected by erratic rainfall patterns as a result of climate change (Gentle and Maraseni, 2012), and The World Bank has estimated that 90% of the world's 1.2 billion people living in extreme poverty depend to some extent on natural forest resources (World Bank, 2004). Therefore, any viable or 'sustainable' solution to global problems requires a consideration of all three dimensions individually as well as through their interrelationships, and trade-offs. Each of the three pillars will now be considered in turn.

Socio-cultural sustainability

Social or socio-cultural sustainability occurs when systems, structures and relationships come together to enable current and future generations to support themselves and create healthy and liveable communities (Polese *et al.*, 2000). As tourist numbers to a specific area rise, their presence is bound to impact the social and cultural fabric of the area. This can be in the form of increased traffic, crime, exploitation of marginalised communities, an influx of migrant workers and changing of customs and norms (Zhuang *et al.*, 2019). This 'social carrying capacity' of a destination is therefore an important consideration in recognising that there is a point at which tourism starts to be viewed negatively by the local population (Getz, 1983).

During the 1980s, as tourism started to spread to new countries in South-East Asia, Africa and the Caribbean, concern over the human and cultural consequences of tourism started to develop. Pressure groups such as the UK-based Tourism Concern and Ecotourism Society in the USA were formed to promote responsible tourism, and address concerns related to human rights violations and access to land and resources for local people (Holden, 2016).

Economic sustainability

In relation to the economic pillar, tourism is a significant revenue generator globally; however, most of the income is earned by the multinational tour operators, airlines and other large international hospitality conglomerates (Mihalic, 2016). Bramwell *et al.* (2008) argue that not only should the community at the destination be involved in tourism, they should also share in the economic benefits gained from it. Further, economic sustainability should not come at the expense of the other two pillars (socio-cultural and environmental).

Environmental sustainability

The natural environment (environmental pillar) is impacted by tourism and tourists as a result of air and other forms of travel, waste, pollution, land degradation and biodiversity loss among other contributing factors. Much like the concept of the social carrying capacity of a destination discussed earlier, Getz (1983) highlights the importance of considering the physical and ecological carrying capacity of a destination in future planning as well. The physical carrying capacity relates to the maximum use of a tourism resource, before it starts to degrade, while ecological carrying capacity relates to the degradation of the natural environment of the resource (Getz, 1983).

Protection of natural environments and reducing the harm caused by tourists and tourism goes beyond just being 'green' but rather focuses on the need to understand the characteristics of natural resources upon which a tourist destination is based (Liu, 2003). The fact that a natural resource exists does not in itself create tourist or monetary value, rather, as Liu (2003: 464) argues 'the perception of any resource thus does not rely on its physical properties, but on a range of economic, technological and psychological factors. Resources are not, they become.' Moreover, tourism is uniquely positioned as an industry that can contribute positively through the protection and rehabilitation of the natural environment, and not only impact it adversely.

For example, this can be pursued through the justification and funding for the preservation of flora, fauna and other natural resources; improving the quality of the local environment as an incentive to attract more visitors, and; increasing awareness of the natural resources (Cheung and Li, 2019).

Thus, acknowledgement and an understanding of the interplay between the three pillars of sustainability in shaping the future of the industry are required. This interplay is the underpinning of the preferred definition of sustainable development presented by the World Tourism Organisation (UNWTO, 2005):

Sustainable tourism development meets the needs of present tourists and host regions while protecting and enhancing opportunities for the future. It is envisaged as leading to management of all resources in such a way that economic, social and aesthetic needs can be fulfilled while maintaining cultural integrity, essential ecological processes, biological diversity and life support systems.

Responsible tourism – whose responsibility?

If sustainability is the goal, then responsible tourism is how to get there. It is about the decisions and actions people take to make tourism more sustainable in the long run (Goodwin, 2011). Tourism can be made more sustainable through innovation and disruption, such as the shared economy, slow travel, eco-tourism and numerous other technological solutions. However, a more concerted effort is required to achieve inter- and intra-generational benefits from tourism and limit negative consequences.

Tourism is widely acknowledged as a growing sector of the global economy and contributes to employment, economic growth, environmental protection and poverty alleviation. If managed well, it can preserve the natural and cultural assets on which it depends, through providing the economic means to keep culture and natural assets alive, the protection of those assets and increased visitor appreciation of those assets. Tourism can also empower host communities, generate trade opportunities, and foster peace and intercultural understanding today and into the future (UNWTO and UNDP, 2017).

However, despite these positive effects of tourism, it also has the potential to increase greenhouse gas emissions, create economic leakages, require resource management and has negative impacts on local communities and cultural assets (Goodwin, 2011; Mihalic, 2016). It therefore becomes imperative that the various stakeholders in the tourism industry play an active and collective role in mitigating these negative consequences.

The tourism industry globally and in any given country is made up of a complex web of actors who have a vested interest in the success of sector and individual ventures from different, and sometimes competing, vantage points. These stakeholders can include, *inter alia*, tour operators, the local community, tourists, hotels, airlines, technology companies, different levels of government, local businesses, residents, activist groups and tour company employees (Sautter and Leisen, 1999).

As an example of competing stakeholder vantage points, a study by Morrison and Pickering (2013), conducted into the impact of climate change on Australian ski-tourism, found that although the various stakeholders agreed on the impacts of climate change on the industry (mostly a shorter ski season), the implications and proposed solutions varied considerably. Given that non-snow-based tourism currently accounts for only 30% of revenue, proposed solutions focused primarily on diversification and snowmaking. For the stakeholders whose revenues are directly linked with the snow season, snowmaking was the preferred alternative, however, those not in favour of that option raised concerns about the intensive use of water and its implications for agriculture, fire protection and local ecosystems. This example not only highlights the complexity and competing interests of stakeholders, but also raises policy questions for the governance of the sector.

Empirical evidence from the hospitality sector has also shown that environmental performance is of far lower priority than economic performance. Consequently, the sustainability agenda is more easily accepted by the public rather than private stakeholders, where the focus has traditionally been on economic goals (Mihalic, 2016).

Therefore, on the question of whose responsibility sustainable tourism ultimately is, it is important to add to whom and for what are they responsible? According to Goodwin (2011), responsible tourism requires action on the part of producers, tourists and all stakeholders to make decisions and demonstrate action on sustainability solutions from the originating markets to the destination. This is considered a systems approach to responsible tourism.

A long-term systems approach to responsible tourism

Sustainable tourism development globally faces a number of 'wicked problems', which Rittel and Weber (1973) in their seminal work on the topic describe as problems that are complex, have no final solution, and are characterised by being socially complex. Further, they can be unstructured and difficult to clearly define; ever-evolving; cross-cutting; intricately connected to other problems and issues; and essentially unique (Scherrer and Doohan, 2014: 1004).

In this regard, Phi (2019: 2) is critical of the role of the UNWTO and other industry players based on their oversimplification of problems such as overtourism and sustainable tourism development, which they contend can be 'addressed by management solutions and individual agency'. Phi (2019) argues for wider acknowledgement that natural, economic and socio-cultural systems are complex, interdependent and likely to change in the future. The recognition that tourism represents a complex, rather than simple system is a good start, where a large range of activities and stakeholders need to be managed simultaneously (Schianetz and Kavanagh, 2008). Or as Choi et al. (2017: 3) state, 'the tourism system refers to a functional structure comprising supply and demand components that influence one another'. Chapter 12 discusses systems thinking in further detail.

A growing number of studies in the tourism field are taking a systems lens to investigate phenomena, including threats to tourism sustainability (Mai and Smith, 2015), interaction among and with tourism stakeholders (Morrison and Pickering, 2013), tourism futures (Walker et al., 1998), ecotourism systems (Choi et al., 2017) and understanding destination image (Tegegne et al., 2018). In their study, Mai and Smith (2015) used Butler's (1980) Tourism destination life cycle to explore the relationship between tourism development theory and systems thinking theory to establish a link between system behaviour and structure. They argue that tourism development (behaviour) should focus less on growth and more on sustainability, since growth is almost always impeded by the destination's physical, ecological and social carrying capacity (structure) and can lead to unintended negative consequences in the future. One of the key benefits of systems thinking over traditional analysis is that it focuses on how elements of the system interact with each other over time, rather than analysing the individual elements on their own.

As tourism is expected to grow globally, the task is not to limit the growth, but to manage it in a way that benefits tourists, the destination environment and host populations (Liu, 2003). To achieve this, the future of responsible tourism requires decision-makers to take a systems approach; one where the interconnectedness and complexity of the system is accounted for, and stakeholders are actively involved in each part of the system.

Conclusion

This chapter has shown that the future and very existence of tourism is likely to be impacted by inter- and intra-generational challenges posed by the natural environment, societal development and economic considerations. Therefore, the current growth and innovation in the industry cannot be expected to continue infinitely as we continue to hit planetary boundaries and limits to growth. Rather, this requires an urgent and critical assessment of the entire tourism system and the extent to which the various components and stakeholders in the system are contributing positively or negatively to future generations' ability to share in the benefits from tourism.

Taking this view of tourism within a broader context requires the responsible management of change and transitions. Responsible tourism, which has been framed in this chapter as the actions required to transform and make the tourism industry more sustainable, therefore becomes the responsibility of all stakeholders to reduce the negative and enhance the positive effects of tourism.

Case study: iSimangaliso Wetland Park, South Africa

The iSimangaliso Wetland Park (iSWP) is South Africa's first UNESCO World Heritage Site which promotes conservation and development, as mandated by South Africa's former president and Nobel Peace Prize laureate, Nelson Mandela. Located on the east coast of South Africa, the area spans 280 kilometres of coastline and is made up of 1,328,901 hectares of natural ecosystems. On their website, iSWP state that the park was 'Born out of partnership', with stakeholder representation from local communities, neighbouring countries Mozambique and Swaziland, tourism operators, accommodation providers and government agencies (iSimangaliso, 2019b).

The park was declared a World Heritage Site in 1999 based on its unique beauty, rich biodiversity and natural beauty. Among what iSWP describes as its 'ten jewels', it is home to three major lake systems, beaches, game viewing, coral reefs and other water activities. These features have made iSWP one of the prime tourist destinations in South Africa for both domestic and international tourists (iSimangaliso, 2019a).

The iSimangaliso Authority aims to grow the current tourism market through attracting investment into the hotels and lodges in the area, and identifying opportunities for more ecotourism activities such as game drives, underwater activities and bird watching (Zaloumis, 2007). It is expected that this growth in tourist numbers will create jobs, stimulate economic growth and 'generate revenues that will reduce conservation's reliance on the government's fiscus' while preventing 'over-exploitation of the area and ensure the universal values of the Wetlands are enhanced' (iSimangaliso, 2019b).

The approach to conservation at iSWP can be described as Integrated Conservation and Development (ICD), which aims to reduce poverty and improve human rights and democracy while achieving more efficient conservation outcomes. The rationale for this, according to Dahlberg and Burlando (2009: 37), is so that 'economic loss to local communities caused by restricted access to natural resources in protected areas should be compensated through alternative income sources, thereby reducing dependence on these resources and increasing awareness of conservation benefits.'

Conservation efforts at iSWP depend on the communities around the park, whose future and preservation of cultural heritage depends on the economic returns from iSWP. According to Zaloumis (2007), since the Park Authority was established in 2000, local people have been given priority in business opportunities, jobs, training and development projects. The involvement of local people is in line with the vision of both the government and the local communities that seek to promote ecotourism and nature conservation.

Despite iSWP's economic success, the lived reality for people around the park remains one of exclusion in post-apartheid South Africa (Hansen *et al.*, 2015). The assumed 'win-win' approach to conservation for development at iSWP and globally has raised concerns about the inherently political process and competing social outcomes from tourism to these areas. Dahlberg and Burlando (2009) are critical of ICD projects because they are often too limited, externally owned, have inappropriate funding models and reach a limited base of people. They further argue that there is often poor awareness of the broader social-ecological context, and the benefits promised are often unrealistic and underpinned by superficial participation.

Studies on the success of sustainable tourism development at iSWP have shown mixed results. In one study, Hansen *et al.*, (2015) found that while community participants were involved in the activities at the park, there have been limited advances in their capability development. This was argued as being a result of the fundamentally neoliberal approach to sustainable tourism development at the park, where the market is seen as the best approach to providing for people's welfare. In another study, Fairer-Wessels (2016) argues that sustainable livelihoods are more critical than sustainable tourism development because communities first need to survive. The main reason rural communities around iSWP are unable to break out of the cycle of poverty is the lack of training and education to mobilise themselves economically, including through local tourism development. Finally, using a systems lens, Stoffelen *et al.*, (2016) found that despite national policies that present a sound rationale through ICD, there is often an implementation gap on the ground in relation to the ability of local communities to actively participate in the activities at the park, and the disconnect between their reality and the national and international institutions (e.g. The World Bank) that espouse sustainable development.

This case has shown that the best laid plans for sustainable development may achieve some, but not all, of their intended objectives.

Discussion questions

1 Describe and discuss the relevant UN Sustainable Development Goals in relation to this case study.

2 To what extent do you agree with the critique of 'Conservation for Development' in the case of the iSimangaliso Wetland Park?

References

Bramwell, B., Lane, B., McCabe, S., Mosedale, J. and Scarles, C. (2008) Research perspectives on responsible tourism, *Journal of Sustainable Tourism*, **16** (3), 253-257.

Butler, R.W. (1980) The concept of a tourist area cycle of evolution: implications for management of resources, *Canadian Geographer/Le Géographe canadien*, **24** (1), 5-12.

Cheung, K. S. and Li, L. H. (2019) Understanding visitor–resident relations in overtourism: developing resilience for sustainable tourism, *Journal of Sustainable Tourism*, **27** (8), 1197-1216.

Choi, Y., Doh, M., Park, S. and Chon, J. (2017) Transformation planning of ecotourism systems to invigorate responsible tourism, *Sustainability*, **9** (12), 2248.

Dahlberg, A.C. and Burlando, C. (2009) Addressing trade-offs: Experiences from conservation and development initiatives in the Mkuze Wetlands, South Africa, *Ecology and Society*, **14** (12), 36-48.

Fairer-Wessels, F.A. (2016) Determining the impact of information on rural livelihoods and sustainable tourism development near protected areas in Kwa-Zulu Natal, South Africa, *Journal of Sustainable Tourism*, **25** (1), 10-25.

Gentle, P. and Maraseni, T. N. (2012) Climate change, poverty and livelihoods: adaptation practices by rural mountain communities in Nepal, *Environmental Science & Policy*, **21**, 24-34.

Getz, D. (1983) Capacity to absorb tourism: Concepts and implications for strategic planning, *Annals of Tourism Research*, **10** (2), 239-263.

Goodwin, H. (2011) *Taking Responsibility for Tourism*, Woodeaton, Oxford: Goodfellow Publishers.

Hansen, M., Faran, T. and O'Byrne, D. (2015) The best laid plans: using the capability approach to assess neoliberal conservation in South Africa – The case of the iSimangaliso Wetland Park, *Journal of Environment & Development* **24** (4), 395-417.

Holden, A. (2016) *Environment and Tourism*, London: Routledge.

iSimangaliso (2019a) South Africa's first World Heritage Site, *iSimangaliso.com*, n.d., viewed 13/10/2019, https://isimangaliso.com/about-us/south-africas-first-world-heritage-site/

iSimangaliso (2019b) Born out of partnership, *iSimangaliso.com*, n.d., viewed 13/10/2019, https://isimangaliso.com/about-us/born-out-of-partnership/

Liu, Z. (2003) Sustainable tourism development: A critique, *Journal of Sustainable Tourism,* **11** (6), 459-475.

Mai, T. and Smith, C. (2015) Addressing the threats to tourism sustainability using systems thinking: a case study of Cat Ba Island, Vietnam, *Journal of Sustainable Tourism,* **23** (10), 1504-1528.

Mihalic, T. (2016) Sustainable-responsible tourism discourse - Towards 'responsustable' tourism, *Journal of Cleaner Production,* **111** (Jan), 461-470.

Morrison, C. and Pickering, C.M. (2013) Perceptions of climate change impacts, adaptation and limits to adaption in the Australian Alps: the ski-tourism industry and key stakeholders, *Journal of Sustainable Tourism,* **21** (2), 173-191.

Phi, G.T. (2019) Framing overtourism: a critical news media analysis, *Current Issues in Tourism,* 1-5.

Polese, M., Stren, R.E. and Stren, R. eds. (2000) *The Social Sustainability of Cities: Diversity and the management of change,* Toronto: University of Toronto Press.

Rittel, H.W. and Webber, M.M. (1973) Dilemmas in a general theory of planning', *Policy Sciences,* **4** (2), 155-169.

Sautter, E. T. and Leisen, B. (1999) Managing stakeholders a tourismplanning model, *Annals of Tourism Research,* **26** (2), 312-328.

Scherrer, P. and Doohan, K. (2014) Taming wicked problems: towards a resolution of tourism access to Traditional Owner lands in the Kimberley region, Australia, *Journal of Sustainable Tourism,* **22** (7), 1003-1022.

Schianetz, K. and Kavanagh, L. (2008) Sustainability indicators for tourism destinations: A complex adaptive systems approach using systemic indicator systems, *Journal of Sustainable Tourism,* **16** (6), 601-628.

Stoffelen, A., Adiyia, B., Vanneste, D. and Kotze, N. (2019) Post-apartheid local sustainable development through tourism: an analysis of policy perceptions among 'responsible' tourism stakeholders around Pilanesberg National Park, South Africa, *Journal of Sustainable Tourism,* **28** (3), 1-19.

Tegegne, W. A., Moyle, B. D. and Becken, S. (2018) A qualitative system dynamics approach to understanding destination image, *Journal of Destination Marketing & Management,* **8** (Jun), 14-22.

UN (2019) Sustainable Development Goals, *United Nations Sustainable Development Goals Knowledge Platform,* viewed 2/11/2019, https://sustainabledevelopment.un.org/?menu=1300

UNESCO. (2017) UNESCO and the International Year of Sustainable Tourism, *United Nations Educational, Scientific and Cultural Organisation,* viewed 2/11/2019, https://en.unesco.org/iyst4d

UNWTO. (2005) *Making Tourism More Sustainable: A guide for policy makers,* Madrid: United Nations World Tourism Organisation.

UNWTO. (2019) *International Tourism Highlights, 2019 Edition*, United Nations World Tourism Organisation, Madrid, viewed 10/11/2019, https://doi.org/10.18111/9789284421152

UNWTO. (2020e) *International tourism growth continues to outpace the global economy*, January 2020, viewed 15/07/2020, https://www.unwto.org/international-tourism-growth-continues-to-outpace-the-economy

UNWTO and UNDP. (2017) *Tourism and the Sustainable Development Goals – Journey to 2030*, Madrid: United Nations World Tourism Organisation.

Walker, P.A., Greiner, R., McDonald, D. and Lyne, V., (1998) The Tourism Futures Simulator: A systems thinking approach, *Environmental Modelling & Software*, **14** (1), 59-67.

WCED. (1987) *Our Common Future*, Oxford: Oxford University Press.

World Bank. (2004) *Sustaining Forests: A development strategy*. Washington DC: The International Bank for Reconstruction and Development/The World Bank.

Zaloumis, A. (2007) A name for a miracle, *Isimangaliso News*, **1** (1).

Zhuang, X., Yao, Y. and Li, J. (2019) Sociocultural impacts of tourism on residents of World Cultural Heritage Sites in China, *Sustainability*, **11** (3), 840-859.

11 Future Proofing a Crisis

Introduction

The nature of the international tourism industry makes it prone to crises. Its service characteristics together with the number of potential external threats considered beyond its control, contribute to its susceptibility and make the risks difficult to manage (Evans and Elphick, 2005; Santana, 2008; Pforr, 2009; Nian *et al.*, 2019). Interruptions to services at the destination itself and within transit routes, as well as the (mis)perceptions of consumers in distant markets, contribute to an increased vulnerability to the short and longer term effects of a crisis

The UN World Tourism Organisation (UNWTO) considers a crisis, in the context of the travel and tourism industry, as 'any unexpected event that affects traveller confidence in a destination and interferes with the ability to continue operating normally' (2011). The terms 'crisis' and 'disaster' are often used interchangeably within the literature although it may be argued a difference in meaning exists (Rindrasih *et al.*, 2019). The scale and responsibility, or control factors within the causation of the event, appear to be the key elements used to distinguish the two terms. Faulkner (2001: 136), for example, makes the distinction between the two terms by defining a crisis as 'a self-inflicted event caused by problems, such as inept management structures and practices or a failure to adapt to changes'; and a disaster as 'a situation in which an enterprise (or group of enterprises) is confronted with sudden unpredictable and catastrophic changes over which it has little control'.

Within a tourism context, Bierman (2016) makes the distinction of crises as being either Category 1 or 2. Category 1 crises are beyond the control of management and include natural disasters, acts of terrorism and war, crime, political conflicts and sudden economic downturns. Category 2 crises result from management's failure to act or implement processes to prepare for or deal with predictable risks such as high staff turnover, lack of insurance in a situation of fire or flood, service and equipment failure, financial fraud and loss of data (Bierman, 2016).

For the purpose of this chapter discussion, a definition adapted from Beirman (2003: 4) relating to a destination crisis will be used, primarily focused on the large-scale nature of the crisis events discussed, these being considered as Category 1, and beyond the control of destination's management and tourism authorities:

> a crisis is a situation requiring radical management action in response to events beyond the internal control of the organisation, necessitating urgent adaption of marketing and operational practices to restore the confidence of employees, associated enterprises and consumers in the viability of the destination.

From a travel and tourism perspective, crisis management refers to the planning and implementation of strategies directed towards managing the negative effects of crises and disasters, particularly on the local tourism industry (COMCEC, 2017). Consequently, crisis management is defined as:

> an ongoing integrated and comprehensive effort that organizations effectively put into place in an attempt to first and foremost understand and prevent crisis, and to effectively manage those that occur, taking into account in each and every step of their planning and training activities, the interest of their stakeholders (Santana 1999: 8).

The output of crisis management planning is the development of crisis management response plans. These aim to build resilience amongst the local community and business operators by equipping them with skills and strategies to cope with a wide range of scenarios. Recognised as a concept in 1962, crisis management incorporates risk analysis and management: disaster planning, prevention, preparedness and response, as well as the overall recovery process, including relationship development with the media (Young and Montgomery, 1998; Granville *et al.*, 2016; Huertas and Oliveita, 2019).

Future crisis management plans must take into account the open system characteristics of the international tourism industry and additional factors required to contend with the increased frequency and magnitude of terror-related crises (Rindrasih *et al.*, 2019). Destination crisis management plans must be flexible, clearly outline decisional roles and responsibilities, utilise effective communication channels and ultimately be integrated into the destination's overall strategic planning processes (Evans and Elphick, 2005).

The impact of crises on the international tourism industry

The impacts experienced by a destination and its tourism industry will vary depending upon the type of crises and its magnitude, resulting in either short-term or long-term implications. Typically, destinations experience physical damage to the environment and tourism infrastructure and a decline in tourist visitation, which subsequently results in economic repercussions for the local community, including decreased spending, a loss of income for local operators and unemployment. Tourism crises tend to also affect the regions surrounding those initially damaged and exert both positive and negative impacts on the area. Kasperson *et al.* (1992) consider these secondary impacts as consequences that spread further than those directly impacted by the original crisis event. For example, during the 1990-1991 Persian Gulf War and the 2004 Boxing Day Indian Ocean Earthquake and Tsunami, where neighbouring countries too were deemed unsafe to visit, these destinations experienced decreased tourism demand (Rindrasih *et al.*, 2019; Evans and Elphick, 2005). Meanwhile, those destinations regarded as alternate tourist destinations experienced increased tourism visitation (Evans and Elphick, 2005). The media often play a role in sensationalising the crises and prolonging the recovery process of not only the immediate destination, but also the surrounding areas (Murphy and Bailey, 1989). Non-factual information about the affected area and distribution of adverse images may be reported, which in turn may contribute to the development of negative perceptions. A positive relationship between the local tourism industry and the media must be developed and media coverage focusing on the recovery of the destination encouraged, including those regions declared safe and business reopening. Investigation into the impact of crises on the performance of the tourism industry is relatively new, with inquiry previously inclined to focus around single events and destinations and little research conducted on their secondary impacts or spill-over effects (Rindrasih *et al.*, 2019). Additionally, the quality of this research varies and remains a challenge for destinations and tourism operators.

The impacts of terrorism on the international tourism industry

Without doubt, many industries are severely disrupted by the impacts of terrorism; however, the effects upon the tourism sector are notably profound (Evans and Elphick, 2005). These types of events are inclined to have the most immediate and greatest impact upon a destination's tourism and significantly reduces overall tourism demand (Drakos and Kutan, 2003; Buigut *et al.*, 2017).

When compared to natural disasters, human-induced disasters or crises, such as terrorist attacks, are inclined to intimidate potential tourists and have a greater negative impact on the destination and recovery process (Buigut *et al.*, 2017; Rindrasih *et al.*, 2019).

Large scale terrorist attacks, such as 9/11, both the 2002 and 2005 Bali Bombings and recent attacks experienced in France, Spain and the United Kingdom clearly indicate that terrorism in particular poses a major threat to the tourism industry and its future (Evans and Elphick, 2005; Sharpley, 2005). Although early acts of terrorism can be traced back to Roman times (Poland, 1988; Schlagheck, 1988), terror related acts began to make headlines during the 1970s and peaked during the mid-1980s, before the more recent events of the past two decades. The 1972 Munich Olympic Games attack along with Tokyo's Narita Airport bombing, the Pan Am 103 bombing over Lockerbie in Scotland and an attack on a tourist bus in Cairo were some of those most highly publicised of 22 terrorism-related attacks occurring between 1972 and 1988 (Mickolus, 1980; Fletcher, 1993; Sonmez and Graefe, 1998). Furthermore, the World Trade Organisation attributed a loss of $105 billion in tourism receipts to terrorism in 1986 (Sonmez and Graefe, 1998a). Terrorism media headlines continued throughout the 1990s and as Jenkins (1988) predicted, the world bore witness to increased future terrorism via the use of weapons of mass destruction. Although terrorism-related crises are not a new phenomenon in tourism management:

> ...their nature and magnitude are not easily comparable to those of past years' events and the overall threat of terrorism remains very serious and one of growing concern... (US Department of State, 2013:1, in Baker, 2014).

Terrorism may be defined as:

> ...the premeditated use or threat of use of extra-normal violence or brutality by sub-national groups to obtain a political, religious or ideological objective through intimidation of a huge audience, usually not directly involved with the policy making that the terrorists seek to influence (Enders and Sandler, 2002: 145-146).

Terrorism and tourism are paradoxically linked due to their shared characteristics in which both cross international borders, generally involve citizens of differing countries, and both are inclined to employ travel and communications technologies (Schlagheck, 1988; Baker, 2014). Western tourists are often deliberately targeted by acts of terrorism as they signify or are viewed as ambassadors for their countries; they represent a difference in socio-cultural values and their involvement assures media coverage worldwide (Weimann and Winn, 1994; Sonmez, 1998). Due to the nature of terrorist acts being unpredictable in terms of geographical location, timing and scale, destinations and tourism operators face numerous challenges (Evans and Elphick, 2005). Such events are extremely difficult to foresee and as a result, planning for such events is difficult. Destination Management Organisations (DMOs) and operators do their best to assess and identify risks a destination may be prone to and attempt to mitigate such risks.

There is no doubt that the events of September 11 in New York City in 2001 had a detrimental and long-lasting impact on the tourism industry, not just within the United States, but with major global repercussions. The World Travel and Tourism Council (WTTC) issued an economic and employment implication projection of a 10% reduction in tourism less than one month following the terrorist attack (Beirman, 2003). Security intensified and travel warnings were issued by most governments in the immediate aftermath, deterring their citizens from travelling to the Middle East and South Central Asia regions. Some prospective travellers interpreted these warnings as a sign to avoid travelling to Europe, Asia and the USA. While many travellers were immediately reluctant to travel internationally and cancelled travel plans, others redirected their travel focus and opted for alternate, less high profile destinations outside major cities, or domestic travel instead (Beirman, 2003; Chen and Noriega, 2004). The World Tourism Organisation (UNWTO 2001) reported that while the USA, Middle East and parts of South East Asia experienced a decline in forward bookings, especially from Australians and New Zealanders, a strong growth in demand for travel to South America, the South-West Pacific and domestic destinations was reported.

It is difficult to judge the real risk of terrorism and often the tourist's fear of terrorism is not relative to the actual level of risk (Baker, 2014). Although there is a relatively low risk or chance of being affected by terrorism, the threat for tourists is very real which will influence their decision to travel (Sonmez and Graefe, 1998; Kozak et al., 2007). For some destinations, persistent terrorism tarnishes the destination's positive image and even jeopardises its entire long-term tourism business. 'The greatest impact on tourist demand comes from terrorist attacks where tourists and locals are the direct target or victims of the attack' (Baker, 2014: 64).

The impact of natural disasters on the international tourism industry

In addition to threatening lives and the livelihood of the local community, natural disasters characteristically damage the natural environment, tourism industry resources and service facilities, as well as often undermine the future tourism industry market (Nian *et al.*, 2019). Examples of the repercussions for international tourism following the 2004 Indian Ocean Tsunami, Hurricane Katrina in 2005, the 2019 White Island volcanic eruption and the 2020 Australian bushfires are discussed below.

The 2004 Indian Ocean Tsunami is recorded as being one of the most catastrophic natural disasters to take place in modern history, demolishing Indonesia's Aceh Province, along with the west coast of both Malaysia and Thailand and the east coast of India and Sri Lanka. The east coast of Africa was also affected (Sharpley, 2005; Rindrasih *et al.*, 2019). Reports indicate that approximately 230,000 people lost their lives in over 14 countries (Langill, 2018). Key tourist regions in Thailand, including Phuket and beach resorts along the Andaman Sea, experienced an instant decrease in tourism immediately following the disaster. Despite the local government assisting local hotel and tour operators to reopen for business shortly afterwards, the following year showed a decline in tourism revenue by 71% in the Phuket region (Birkland *et al.*, 2006). This downturn in tourism remained constant over the next two years, and it was only when these former tourist regions rebuilt hotels and tourism related facilities that were installed with new tsunami warning systems that a sense of safety was renewed and tourists began to return. Fortunately for Thailand and many of its affected tourist regions, the impact on the local tourism industry was relatively short-term.

Not all natural disaster affected tourist regions recover so quickly. Prior to Hurricane Katrina in 2005, the city of New Orleans generated US$4.9 billion in tourism and attracted 10.7 million visitors annually. However, during 2006 the city only received 3.7 million tourists who spent US$2.8 billion (New Orleans Tourism Marketing Corporation, 2015). Despite the number of visitors having rebounded to 9.5 million generating a record US$6.8 billion in 2014, visitor numbers had not returned to previous levels. From a tourism perspective, the city essentially lost a decade of tourism visitation and income and is still dealing with the image of being a 'damaged destination' (Elliott, 2015).

New Zealand's volcanic White Island unexpectedly erupted explosively on the 9th December 2019, killing 21 and badly injuring a number of tourists (Reid, 2020). Public tours have been conducted to the privately owned island for over 30 years with an estimated 10,000 tourists visiting each year (Shrimpton, 2019). All tours to White Island were cancelled immediately following this tragedy and the tour operators have no immediate plans to resume their tourist trips until further investigation into the explosion has occurred. Criticism surrounds White Island volcanic alert system which was raised from 1 to 2 on a scale of 5 three weeks prior to the crisis incident, indicating 'moderate to heightened unrest' with potential for eruption (Morton,

2019). While this rating level was still considered safe for travel operators to access the island, these alert system ratings certainly do not forecast eruptions deemed 'inherently unpredictable' and a degree of risk will always be present (Andrews, 2019).

In January 2020 the east coast of Australia experienced its worst bushfires since 2009 with an estimated land area of 5.9 million hectares being burnt, equivalent to the European countries of Denmark and the Netherlands combined (White and Gilbert, 2020). The true impact of these catastrophic bushfires on the local tourism industry is yet to be realised, however reports from the Australian Tourism Export Council (ATEC) have indicated the impact on the Australian tourism industry is expected to be as great as AU$4.5 billion (Carruthers, 2020). Many neighbouring local tourist regions have suffered as a result of the bushfire crisis, and these safe and unaffected areas are desperately encouraging tourist visitation and promoting themselves as 'open for business'. Generous worldwide support, in addition to government funding, has been provided to assist those most in need; however, reports of those victims not being able to access this funding raises the question of how such situations should be better managed in the future.

The impact of disease outbreaks on the international tourism industry

Over the last ten years there have been five declarations of international public health emergencies: the 2009 swine flu pandemic; a polio outbreak and the Western Africa Ebola outbreak in 2014; the 2015 Zika virus outbreak and another Ebola outbreak in the Democratic Republic of the Congo in 2019 (Reynolds, 2020). Prior to this in 2003, severe acute respiratory syndrome (SARS) swept through most of Asia infecting more than 8,000 people and killing 774. The World Health Organisation (WHO) declared the most recent novel coronavirus outbreak (Covid-19) a global health emergency on 31st January 2020. Six weeks later on 12th March 2020, the WHO upgraded this declaration to a global pandemic due to the rapid rise in cases outside of China affecting a growing number of countries. Similar to SARS, COVID-19 is thought to have originated in China from wild animals sold in markets; however it took less than two months to infect 75% of the number of people infected by SARS over a nine-month period (Hollingsworth, 2020). At the time of writing, almost 4,650 deaths and an estimated 84,778 cases of the infection have been reported in China with at least another 25 countries having confirmed cases (News China, 2020). Since the initial outbreak of the disease in China, higher confirmed cases and death rates have been recorded in the European nations of Spain (252,513 cases and 28,396 deaths), Italy (242,149 cases and 34,914 deaths) and the United Kingdom (286,983 cases and 44,517 deaths) while the USA (2,973,695 cases and 130,893 deaths) and Brazil (1,668,589 cases and 66,741 deaths) figures continue to climb. To date, global figures stand at over 11.8 million Covid-19 cases and 545,481 deaths (WHO, 2020).

It is evident that many countries have learnt from the SARS crisis in 2003 and as a result were quick to impose precautionary protocols in an attempt to mitigate any further spread of Covid-19. What initially began with some countries temporarily banning arrivals from mainland China and some cruise liners with infected passengers being denied entry to a number of ports, closures of some international borders and strict 14 day quarantine rules have been enforced upon entry into many other countries. In some cases, airlines have grounded up to 90% of their international flights and stood down as many as 20,000 staff members and crew as a result of restricted international travel demand (The Guardian, 2020). A number of countries have 'locked down' their societies, only allowing their population to leave their homes to access essential services such as supermarkets, medical support and some places of employment such as childcare centres and hospitals. As the economic repercussions of this pandemic are predicted to overtake that of the SARS crisis, the true impact of this health crisis remains to be seen. As the world's largest outbound tourism-related spender, China represents one fifth of international tourism spending; therefore, the Covid-19 pandemic is expected to impact the international tourism industry tremendously (UNWTO, 2019).

Crises take many forms, occur randomly and potentially may have a major impact on both the local and international tourism industry. Recovery processes and their duration will also vary depending on the type and severity of the crises, and those destinations which have experienced a situation of crisis before are likely to have a crisis management plan in place, having learned from their past experience.

Crisis management models and preparedness strategies

Various models or frameworks exist in relation to preventing, managing and minimising the impact of a crisis or disaster upon the local destination and tourism industry (Faulkner, 2001; Cronstedt, 2002; Ritchie, 2004; Laws and Prideaux, 2006; Hystad and Keller, 2008; COMCEC, 2017). Many of these crisis or disaster management models involve a multi-phase approach which includes a series of sequential steps or stages from initial risk assessment, prevention and preparedness through to the crisis response strategy and facilitating the recovery process. The final stage of most models involves an evaluation of the specific crisis management process and identification of areas for future improvement. Faulkner (2001) proposes a six stage framework for understanding the various steps of a destination crisis: Pre-event; Prodromal; Emergency; Intermediate; Recovery and; Resolution. Ritchie's (2004) Crisis and Disaster Management Model, which builds upon Faulkner's (2001) model, consists of a three step approach to managing the recovery and restoration of a destination in crisis: prevention and planning; strategic implementation and; resolution, evaluation and feedback. Both of these crisis management models are considered from a destination perspective.

Many destinations use past crisis events as motivation for action and as examples for future crisis management planning. Despite a wide range of crisis and disaster management strategies being available, the preparedness levels within the tourism industry historically remains low (Faulkner and Vikulov, 2001; Hystad and Keller, 2008; Paraskevas et al., 2013). Primary reasons for this viewpoint are attributed to beliefs that disasters are unlikely to occur, a lack of funding, time, and expertise in crisis management (Faulkner and Vikulov, 2001; Hystad and Keller, 2008).

Evans and Elphick (2005) evaluated a number of crisis management models in regard to their strategic planning value within the international tourism industry. These researchers concluded that the travel sector must: prepare contingency plans; define decisional and informational roles and responsibilities; maintain a level of flexibility; and, ensure an evaluation is conducted once the immediate crisis is over to review and allow for future improvement within the management process (Evans and Elphick, 2005: 148-149). Mair *et al.* (2016), in their review of 64 articles concerning post-disaster and post-crisis recovery for tourist destinations, found the key themes to emerge from the literature related to media sensationalism including a lack of communication amongst stakeholders; the value of effective marketing message selection; lack of disaster-management plans; destination image and reputation loss; and tourist behaviour changes following a crisis and/or disaster. Effective and timely responses, relationship marketing and destination image restoration were considered important in effective destination crisis management (Mair *et al.*, 2016).

Future approaches to destination crisis management

Within the field of crisis management, a strong belief exists that crises are changing shape and becoming more complex in nature (Rosenthal, 1998; Kouzmin and Haynes, 1999; Rosenthal *et al.*, 2001). Crises are becoming increasingly transboundary and interconnected due to globalisation, increased mass communication, social fragmentation and changes to state authority (Boin and Lagadec, 2000). Likewise, disasters have changed in nature and magnitude, especially in relation to terror-related crises (Department of Foreign Affairs and Trade, 2019).

Rindrasih *et al.* (2019) suggest that a *systems thinking* approach, as outlined in more detail in Chapter 12, could be applied to future crisis management in order to understand the complexities associated with the international tourism industry. The authors argue that the nature of the industry, being considered an 'open' system, is more suitable to be viewed and studied as a complex system in order to understand the impacts and relationships within the industry (Rindrasih *et al.*, 2019). Others in the field agree a more proactive and flexible approach whereby crisis management is integrated into strategic planning processes, rather than the rigid sequential orderly nature of many crisis management models and frameworks, may enable a more effective destination response to be achieved (Evans and Elphick, 2005; Martens *et al.*, 2016).

Similarly, lessons and implications for future crisis management and policy development may be considered using elements of 'surprise management' theory. Surprise management emphasises adaptability, collaboration and citizen engagement, while drawing on chaos and complexity theories in order to manage situations of hyper-uncertainty (Farazmand, 2007). This management theory is a long-term approach requiring unrestrained resources, cutting-edge knowledge skills, leadership and disciplined authority to develop a crisis management strategy which anticipates all conceivable and inconceivable future circumstances. Subsequently, such a crisis management approach is expensive to develop and maintain but it may be argued that this approach is 'a national asset with no substitute' (Farazmand, 2007: 157). Surprise management is credited with the enormously successful response to the 2004 earthquake crisis in Iran, which killed 50,000 of the 80,000 residents of Bam. In less than 24 hours, chaos had been contained and international response teams found themselves redundant upon their arrival (Farazmand, 2007). Acknowledging that not all destinations are in the position to adopt a complete surprise management approach, its fundamentals may be considered and incorporated into overall destination management and crisis planning.

DMOs play a key role within the crisis management planning process due to their coordination role in marketing related activities, including the quest for destination competitiveness (Granville *et al.*, 2016). Gaining a better understanding of the consumer response to disastrous events also plays an important role in minimising negative destination image (Walters *et al.*, 2015). The DMO is particularly instrumental as a leader and facilitator in relationship development and communication amongst destination stakeholders. Digital dissemination of critical information from emerging digital platforms is required throughout the various stages of a crisis in order to effectively alert, inform and assist in the recovery process (Granville *et al.*, 2016). Key to reassuring public safety and minimising negative destination image perception is effective communication (Beirman, 2003; Ritchie, 2004). Destinations need to focus on increasing their ability to effectively alert and inform the public in the event of a crisis. In Melbourne, Australia, following the tragic 2017 attack on pedestrians along a popular central city shopping strip on Bourke Street, terror alert systems, including CCTV and speakers along with cement barriers were installed across the central business district area. These emergency warning systems are designed to inform the public in an emergency and instruct them to move towards a specified safe location or to take shelter nearby. These systems are intended to be used in conjunction with text messages and alerts via social media platforms, with the idea being to ensure the public have the appropriate information required to manage a crisis incident, rather than incite a state of panic (ABC News, 2017).

Similarly, some destinations prone to earthquakes and tsunamis have installed early warning systems, particularly in dense tourist areas. Following the 2006 Indian Ocean Tsunami, 19 warning towers were installed in the Phuket province. These are designed to send a loud warning to tourists across entire villages in the event of an earthquake and often subsequent tsunami (Asia News Network, 2018). It is important that once these sophisticated warning systems are installed, they are routinely tested, maintained and monitored to ensure they remain an effective crisis management tool. Fears of a lack of maintenance in some of the Phuket warning towers were recently raised, with warning volumes in some cases, reported as low as 10% of full capacity (Asia News Network, 2018).

Summary

Future destination crisis management plans, to be effective, must be flexible, demonstrate timely responses, utilise emerging digital media platforms to deliver accurate and up to date information, and integrate into the destination's overall strategic planning processes. A symbiotic relationship exists between the local destination community, government and travel authorities and requires a coordinated approach in preparing for a crisis, including the specification of decisional roles and responsibilities. (Evans and Elphick, 2005). Tourism operators must be willing to engage with crisis preparedness strategies (Walters et al., 2015) and related DMO-coordinated activities, to ensure resilience is attained in the tourism industry (Granville et al., 2016).

A lack of research focusing on the multi-phases of disaster management and the existing relationships within the sector mean there is a limited understanding of the way in which the tourism industry manages change, complexity and subsequent disruptions commonly associated with disasters (Granville et al., 2016). An increased understanding of these areas is required more urgently due to the changing nature and scale of crises, especially those that are natural, health and terror-related.

Case study: The impact of terrorism on France as a popular tourist destination

Introduction

In the years since the 9/11 terrorist attacks on the World Trade Centre in New York, the world has been subjected to an increasing number of terrorism-related incidents. Destinations in Europe, once considered safe, have been selected as prime terrorism targets in more recent years. France has enjoyed the number one position as the most visited destination in the world for over 20 years (Abadi, 2019). This popular tourist haven has experienced at least 12 major terrorism-related events between 2012 and 2018, many of which have specifically targeted tourists. Visitor arrival figures indicate that 89 million international tourists entered the country during 2018, a 3% increase from the previous year and these tourists' direct contribution to France's GDP was US$67 billion (UNWTO, 2019). The travel and tourism industry represents 9.7% of France's GDP, with 30% of tourism revenue gained from international tourists, while the remaining 70% is from domestic tourism. Tourism consumption (both by domestic and international tourists) represents approximately 7.5% of the nation's GDP, with direct and indirect tourism-related employment accounting for over 2 million jobs (OECD, 2018). Tourism is obviously a major contributor to the economy and the threat of increased terrorism, particularly over the last decade, has the potential to jeopardise future visitation.

Key reasons for visitation to France and Paris

Over 89 million international tourists visited France during 2018 (see Figure 11.1) motivated largely by its appealing city lights, romantic image, world class gastronomy, diverse landscape and its array of cultural and heritage sites. Disneyland, the

Louvre Museum, the Eiffel Tower, Chateau de Versailles and Centre Pompidou are amongst the most popular tourist sites visited. The Louvre Museum, featuring the Mona Lisa, drew a record 10.2 million visitors in 2018, more than any other museum in the world (Louvre, 2018).

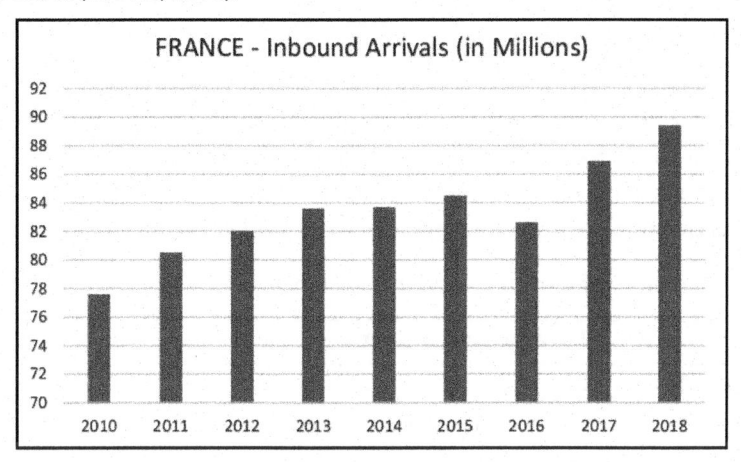

Figure 11.1: Inbound arrivals to France 2010-2018 (millions). *Source:* UNWTO, 2019.

France's diverse landscape of mountains, beaches and rural countryside, including popular winter ski resorts like Chamonix and Val d'Isère, the beaches of the French Rivera and wine regions of Bordeaux, Burgundy and Champagne, appeal to visitors all year round. Approximately 24 million foreign tourists visit Bordeaux, Burgundy and France's other wine regions each year (Regional Tourism Committee, 2019). Culturally, France is home to 39 sites on UNESCO's World Heritage list, placing it equal fourth in the global rankings (UNESCO, 2019). The gastronomic cuisine of the French is also listed on UNESCO's world intangible heritage list, recognising the social culture of the French, including the way in which they gather around a table celebrating the most important moments in their lives (Samuel, 2010).

In addition to its 34 international airports, France is accessible by land from eight different countries. Together with its reliable and easily accessible public transport, this makes it popular amongst international travellers. Approximately 40% of international arrivals originate from the United Kingdom, Germany and Belgium (OECD, 2018). Additionally, it is estimated that between 15 and 20 million of the country's visitors are just passing through en-route to its European neighbours Italy or Spain.

Crisis history summary

Terrorism acts commenced throughout Western Europe during the 1970s with France experiencing a number of attacks during the 1990s, primarily in association with the Armed Islamic Group (GIA) (Rault, 2010). Those nations supporting the Algerian Government, following the 1991 elections, became prime targets by the GIA and other 'home grown' covert Islamic groups. The 1990s attacks demonstrated to the French Government the degree of sophistication that such terrorist groups were capable of (Rault, 2010). Since the 1990s the French population had become accustomed to living under the threat of terrorism, however, the implementation of counterterrorism activities had proved effective in combating any subsequent attacks up until 2012 (Rault, 2010).

Between 2012 and 2018, France experienced over 25 terrorism-related events (see Table 11.1) with 2015 and 2016 undoubtedly two of the worst years. Today's Islamic terrorist groups continue to share a similar ideology to those in the 1990s, which conflicts with democracy and any form of social and political progress (Rault, 2010).

Table 11.1: History of terrorism incidents in France 2012-2018

Date	Location and Incident	Casualties
2012: 11-22 March	Toulouse and Montauban: Shooting	7 killed
2013: 23 May	Paris: Soldier attacked at La Defence	
2014: 20 December	Joue-les-Tours: Knife attack on police officer	1 killed
2014: 21 December	Dijon: Pedestrians ran down	11 injured
2014: 22 December	Nantes: Pedestrians ran down at a Christmas market	10 injured
2015: 7 January	Paris: Charlie Hebdo Office shootings	12 killed
2015: 8 January	Paris: Shooting	5 killed
2015: 3 February	Nice: Soldiers Attacked at Jewish Centre	3 injured
2015: 26 June	Grenoble: Beheading at chemical factory	1 killed
2015: 21 August	Paris: Armed gunman on train averted	3 injured
2015: 13 November	Paris: Series of simultaneous attacks	130 killed
2016: 13 June	Paris: Home Attack on police commander and wife	2 killed
2016: 14 July	Nice: Pedestrians ran down on Bastille Day	85 killed
2016: 26 July	Rouen: Hostages held in a Catholic Church	1 killed
2016: 8 September	Essonne: Anti-terror operation	1 injured
2017: 3 February	Paris: Machete armed man arrested	
2017: 16 March	Paris, France: Letter Bomb	1 injured
2017: 18 March	Paris, France: Police officer attacked at Orly Airport	1 injured
2017: 20 April	Paris: Police officers attacked	
2017: 6 June	Paris: Attack on police officer at Notre Dame Cathedral	
2017: 19 June	Paris: Police vehicle rammed at Champs-Elysees	1 killed
2017: 9 August	Paris: Soldier barracks rammed by a car	6 injured
2017: 1 October	Marseille: Young girls attacked at Saint Charles Station	2 killed
2018: 23 March	Trebes: Supermarket hostages	
2018: 12 May	Paris: Knife attack Palais Garnier	1 killed
2018: 11 December	Strasbourg: Christmas markets attack	3 killed

Source: BBC, 2016; The Telegraph, 2018.

The French government's 'Action Plan against Radicalisation and Terrorism' was introduced in May 2016 and designed to strengthen its predecessor, the 'Action Plan Against Terrorist Networks and Violent Radicalisation', developed against violent radicalisation in 2014. Increased funding and an increased number of action items are amongst key additions, in an attempt to design a global strategy (Uhlmann, 2018). In a response to changing security challenges amidst repeated terrorism incidents between 2016 and 2018, a new action plan was introduced again in 2018, entitled 'Action Plan Against Terror' (General Secretariat for Defence and National Security, 2018). Further adaptations of the 2016 action plan were required due to the changing nature of recent terrorist attacks experienced by the nation, including the formation of a new organisation called 'National Coordination of Intelligence and Counterterrorism', increased human resources dedicated to monitoring potential threats, reinforcement of background checks, a new automated personal data processing system and appointing an ambassador for digital technology responsible for cyber security measures (French Government, 2018). The French government recognises terrorism as an evolving long-term threat to its nation which will require constant monitoring and adaptations to the policies developed in their response to terrorism.

Tourist visitation impact

A 2.2% drop in tourist visitor numbers to metropolitan France in 2016 was reported due to fears of further terror attacks, with concerns primarily amongst Chinese (down 21.5%) and Japanese tourists (down 41.2%) (Hosie, 2017). Europeans appeared less deterred (British down 8.6% and Spanish down 9.9%) with the USA reported as being the least affected (down by 4.9%). This reduction in visitation following the 2015 and 2016 terror incidents was disappointing, but was not as severe as expected (Regional Tourism Committee, 2017). Visitation figures by the end of 2016 resembled those recorded at year end in 2014 and in terms of visitor flows, France maintained their position as the world's most visited country (OECD, 2018). Domestic tourists were recorded as being the first to return, followed by the USA and Chinese tourists. The higher value of the euro currency at the time may also be considered to have influenced visitation, particularly making it more attractive for USA tourists to visit Europe. Not all international markets rebounded as quickly however, with the Japanese tourist market recorded as being slower to return.

The French government allocated 10 million euros to promote France as a tourist destination in 2016. Their aim was to restore a sense of safety and security in the public by implementing tighter security measures around the cities and major tourist attractions, including having police and soldiers visibly patrolling the streets (Schreuer, 2017). Furthermore, tourism initiatives focusing on rural tourism development and bicycle tourism were developed, in the hope of diversifying Frances' tourism offering (OECD, 2018). The short-term tourism impact experienced by France demonstrates that there appears to be a level of acceptance amongst tourists in relation to terrorism risk. Tourists seem to be prepared to live with some level of danger when travelling to popular world cities, such as Paris, and recognise they will never be completely safe (Schreuer, 2017). Terrorism is certainly in travellers' minds but overall, it appears to not be enough to stop them completely from travelling to popular world cities, such as Paris.

Discussion questions

1 Identify factors contributing to France's ability to continually attract large numbers of tourist visitation.

2 From a tourism crisis perspective, what measures does France have in place to protect the nation and its visitors?

3 View France's 2018 'Action Plan Against Terrorism' (http://www.sgdsn.gouv.fr/uploads/2018/10/20181004-plan-d-action-contre-le-terrorisme-anglais.pdf) and identify key counterterrorism initiatives.

4 How do you think terrorism will impact future visitation to France?

References

Abadi, M. (2019) France has been the most-visited country in the world for more than 20 years in a row — but experts think it's about to lose its crown, *Business Insider Australia*, 13 November, viewed 12/12/2019, www.businessinsider.com.au/china-tourism-visitors-france-2018-11

ABC News, (2017) Terror alert speakers, CCTV rolled out across Melbourne's CBD, *ABC News*, 11 December, viewed 14/10/2019, www.abc.net.au/news/2017-12-11/terror-alert-speakers-cctv-rolled-out-across-melbourne-cbd/9245106

Andrews, R. (2019) White Island Volcano: What's going to happen next?, *Forbes*, 11 December, viewed 12/12/2019, www.forbes.com/sites/robinandrews/2019/12/11/white-island-volcano-whats-going-to-happen-next/#615ac6d2ce6e

Asia News Network. (2018) Faulty tsunami warning system in Phuket worries local, *Straits Times*, 5 October, viewed 12/12/2019, https://www.straitstimes.com/asia/se-asia/faulty-tsunami-warning-system-in-phuket-worries-locals

Baker, David Mc. A. (2014) The effects of terrorism on the travel and tourism industry, *International Journal of Religious Tourism and Pilgrimage*, 2 (9), 58-67.

BBC. (2016) Timeline: Attacks in France, *BBC News*, 26 July, viewed 10/10/2019, https://www.bbc.com/news/world-europe-33288542

Bierman, D. (2003) *Restoring Tourism Destinations in Crisis: A Strategic Marketing Approach*, Crows Nest, NSW: Allen & Unwin.

Bierman, D. (2016) *Tourism Risk, Crisis and Recovery Management Guide*, Sydney: David Bierman.

Birkland, T.A., Herabat, P., Little, R.G. and Wallance, W.A. (2006) The impact of the December 2004 Indian Ocean Tsunami on tourism in Thailand, *Earthquake Spectra*, **22** (3), 889-900.

Boin, A. and Lagadec, C. (2000) Preparing for the future: Critical challenges in crisis management, *Journal of Contingencies and Crisis Management*, **8** (4), 185-191.

Buigut, S., Braendle, U. and Sajeewani, D. (2017) Terrorism and travel advisory effects on international tourism, *Asia Pacific Journal of Tourism Research*, **22** (10), 991-1004.

Carruthers, F. (2020) Tourism loses $4.5b to bushfires as overseas visitors cancel, *Australian Financial Review*, 17 January, viewed 14/02/2020, /www.afr.com/companies/tourism/tourism-loses-4-5b-to-bushfires-as-overseas-visitors-cancel-20200116-p53s0s

Chen, R.J.C. and Noriega, P. (2004) The impacts of terrorism: perceptions of faculty and students on safety and security in tourism, *Journal of Tourism and Travel Marketing*, **15** (92-3), 81-97.

COMCEC. (2017) *Risk and Crisis Management in Tourism Sector: Recovery From Crisis in the OIC Member Countries*, COMCEC Coordination Office, August 2017, viewed 21 October 2019, https://www.sbb.gov.tr/wp-content/uploads/2018/11/Ris_and_Crisis_Management_in_Tourism_Sector-.pdf

Cronstedt, M. (2002) Prevention, preparedness, response, recovery – An outdated concept?, *Australian Journal of Emergency Management*, **17** (2), 10-13.

Department of Foreign Affairs and Trade. (2019) Human preparedness and response, *DFAT*, n.d., viewed 5/12/2019, https://www.dfat.gov.au/aid/topics/investment-priorities/building-resilience/humanitarian-preparedness-and-response/Pages/humanitarian-prepraredness-and-response

Drakos, K. and Kutan, A.M. (2003) Regional effects of terrorism on tourism in three Mediterranean countries, *Journal of Conflict Resolution*, **47** (5), 621-641.

Elliot, C. (2015) New Orleans tourism grapples with 'lost decade,' 10 years after Katrina, *Fortune*, 27 August, viewed 12/12/2019, https://fortune.com/2015/08/27/hurricane-katrina-new-orleans-tourism/

Enders, W. and Sandler, T. (2002) Patterns of transnational terrorism, 1970–1999: Alternative time-series estimates, *International Studies Quarterly*, **46** (2), 145–165.

Evans, N. and Elphick, S. (2005) Models of crisis management: An evaluation of their value for strategic planning in the international travel industry, *International Journal of Tourism Research*, **7** (3), 135-150.

Farazmand, A. (2007) Learning from the Katrina Crisis: A global and international perspective with implications for future crisis management, *Public Administration Review*, **67**, 149-159.

Faulkner, B. (2001) Towards a framework for tourism disaster management, *Tourism Management*, **22** (2), 135–147.

Faulkner, B. and Vikulov, S. (2001) Katherine, washed out one day, back on track the next: a post-mortem of a tourism disaster, *Tourism Management*, **22** (4), 331-344.

Fletcher, M. (1993) 'Egypt—Is this the time to visit?', *Travel and Leisure*, **23** (6), 60-64.

General Secretariat for Defence and National Security. (2018) *Action Plan Against Terrorism*, viewed 22/12/2019, http://www.sgdsn.gouv.fr/uploads/2018/10/20181004-plan-d-action-contre-le-terrorisme-anglais.pdf

Granville, F., Mehta, A. and Pike, S. (2016) Destinations, disasters and public relations: Stakeholder engagement in multi-phase disaster management, *Journal of Hospitality and Tourism Management*, **28** (Sep), 73-79.

Hollingsworth, J. (2020) The memory of SARS loom over the Wuhan virus. Here's how the outbreaks compare, *CNN*, 2 February, viewed 10/02/2020, https://edition.cnn.com/2020/01/29/china/sars-wuhan-virus-explainer-intl-hnk-scli/index.html

Hosie, R. (2017) Paris tourist numbers drop due to fears over further terror attacks, *The Independent*, 22 February, viewed 10/09/2019, https://www.independent.co.uk/travel/paris-tourist-numbers-drop-franch-terror-attacks-further-charlie-hebdo-bataclan-shooting-isis-a7592836.html

Huertas, A. and Oliveita, A. (2019) How tourism deals with terrorism from a public relations perspective: A content analysis of communication by Destination Management Organisations in aftermath of 2017 terrorist attacks in Catalonia, *Catalan Journal of Communication and Cultural Studies*, **11** (1), 39-58.

Hystad, P.W. and Keller, P.C. (2008) Towards a destination tourism disaster management framework: Long-term lessons from a forest fire disaster, *Tourism Management*, **29** (1), 151-162.

Jenkins, B. (1988)'Future trends in international terrorism', in R. O. Slater and M. Stohl (eds), *Current Perspectives on International Terrorism*, London: Macmillan, pp. 246-266.

Kasperson, R.E., Golding, D. and Tyler, S. (1992) Social distrust as a factor in siting hazardous facilities and communicating risks, *Journal of Social Issues*, **48** (4), 161-187.

Kouzmin, A. and Haynes, A. (eds) (1999) *Essays in economic Globalization, Transnational Policies and Vulnerability*, Brussels: IOS Press.

Kozak, M., Crotts, J.C. and Law R. (2007) The impact of the perception of risk on international travellers, *International Journal of Tourism Research*, **9**, 233-242.

Langill, M. (2015) Impacts on Thailand's tourism industry after the 2004 Indian Ocean Tsunami, *LaSalle College*, n.d., viewed 5 /12/2019, https://asialasalle2015.wordpress.com/2015/09/17/impacts-on-thailands-tourism-industry-after-2004-indian-ocean-tsunami/

Laws, E. and Prideaux, B. (2006) Crisis management: A suggested typology, *Journal of Travel and Tourism Marketing*, **19** (2-3), 1-8.

Louvre. (2019) 10.2 million visitors to the Louvre in 2018, Louvre, 3 January, viewed 5/12/2019, https://presse.louvre.fr/10-2-million-visitors-to-the-louvre-in-2018/

Mair, J. Ritchie, B.R., and Walters, G. (2016) Towards a research agenda for post-disaster and post-crisis recovery strategies for tourist destinations: a narrative review, *Current Issues in Tourism*, **19** (1), 1-26.

Martens, H.M., Feldesz, K. and Merten, P. (2016) Crisis management in tourism: A literature based approach on the proactive prediction of a crisis and the Implementation of prevention measures, *Athens Journal of Tourism*, **3** (2), 89-102.

Mickolus, E.F. (1980) *Transnational Terrorism - A Chronology of Events, 1968-1979*, Westport, CT: Greenwood Press.

Morton, J. (2019) White Island eruption: Scientists defend volcano warning system amid criticism, *New Zealand Herald*, 11 December, viewed 12/12/2019, https://www.nzherald.co.nz/nz/news/article.cfm?c_id=1&objectid=12292904

Murphy, P.E. and Bailey, R. (1989) Tourism and disaster planning, *Geographical Review*, **79** (1), 36-46.

New Orleans Tourism Marketing Corporation. (2015) Available: https://fortune.com/2015/08/27/hurricane-katrina-new-orleans-tourism/, viewed 5/12/ 2019.

News China (2020) China coronavirus outbreak: All the latest updates. https://www.aljazeera.com/news/2020/02/cloneofchina-coronavirus-outbreak-latest-updates-200213231710117.html, viewed 14/02/2020.

Nian, S., Zhang, J., Zhang, H., Zhang, J., Li, D., Wu, K. , Chen, X. and Yang, L. (2019) Two sides of a coin: A crisis response perspective on tourist community participation in a post disaster environment, *International Journal of Environmental Research and Public Health*, **16** (12), 2073.

OECD (2018) France, in *OECD Tourism Trends and Policies 2018*, Paris: OECD Publishing, pp.169-174.

Paraskevas, A., Altinay, L., McLean, J. and Cooper, C. (2013) Crisis knowledge in tourism: Types, flows and governance, *Annals of Tourism Research*, **41** (Apr), 130-152.

Pforr, C. (2009) Crisis management in tourism: A review of the emergent literature', in C. Pforr and P. Hosie (eds), *Crisis Management in the Tourism Industry: Beating the Odds?*, Surrey: Ashgate, pp. 37-52.

Poland, J. M. (1988) *Understanding Terrorism*, Englewood Cliffs. NJ: Prentice-Hall.

Rault, C. (2010) *French Approach to Counterterrorism,* Vol 3(1), viewed 2/12/2019, https://ctc.usma.edu/the-french-approach-to-counterterrorism/

Regional Tourism Committee (2019). Record breakers: Why France is still the most visited country on earth. Available: https://www.thelocal.fr/20190410/france-retains-crown-as-most-visited-country-on-earth, viewed 6/12/2019.

Reynolds, M. (2020) How China's coronavirus outbreak started, explained. Available: https://www.wired.co.uk/article/china-coronavirus, viewed 9/02/2020.

Reid, N. (2020) White Island tragedy a month on: Landings could be off limits under future tours, *New Zealand Herald*, 9 January, viewed 13/05/2020, https://www.nzherald.co.nz/business/news/article.cfm?c_id=3&objectid=12298888.

Rindrasih, E., Witte, P., Spit, T., and Zoomers, A. (2019) Impact of disaster events on tourism development in Indonesia 1998-2016, *Journal of Service Science and Management*, **12** (02), 93-115.

Ritchie, B.W. (2004) Chaos, crises and disasters: A strategic approach to crisis management in the tourism industry, *Tourism Management*, **25** (6), 669-683.

Rosenthal, U. (1998) Future disasters, future definitions, in E.L. Quarantelli, (ed.), *What is a Disaster? Perspectives on the Question*, London: Routledge, pp.146-160.

Rosenthal, U., Boin, A. and Comfort, L.K. (2001) The changing world of crises and crisis management in U. Rosenthal, A Boin and L.K. Comfort (eds), *Managing Crises: Threats, Dilemmas and Opportunities*, Springfield: Charles. C. Thomas Publisher, pp 6-26.

Samuel, H. (2010) UNESCO declares French cuisine 'world intangible heritage', *The Telegraph*, 16 November, viewed 10/01/2020, https://www.telegraph.co.uk/news/worldnews/europe/france/8138348/UNESCO-declares-French-cuisine-world-intangible-heritage.html

Santana, G. (2008) Crisis management and tourism, *Journal of Travel and Tourism Marketing*, **15** (4), 299-321.

Schlagheck, D. M. (1988) *International Terrorism*. Lexington MA: Lexington Books.

Schreuer, M. (2017) Paris tourism has recovered from 2015 attacks, officials say, *New York Times*, 14 April, viewed 20/12/2019, https://www.nytimes.com/2017/04/14/world/europe/paris-tourism.html

Sharpley, R. (2005) The tsunami and tourism: A comment, *Current Issues in Tourism*, **8** (4), 344-349.

Shrimpton, W. (2019) White Island's long history of unrest hasn't stopped tourism, *Newshub*, 10 December, viewed 10/04/2020, https://www.newshub.co.nz/home/new-zealand/2019/12/white-island-s-long-history-of-unrest-hasn-t-stopped-tourism.html

Sonmez, S. (1998) Tourism, terrorism, and political instability, *Annals of Tourism Research*, **25** (2), 416–456.

Sonmez, S. and Graefe, A.R. (1998) Influence of terrorism risk on foreign tourism decisions, *Annals of Tourism Research*, **25** (1), 112-144.

Sonmez, S. and Graefe, S. (1998a) Determining future travel behavior from past travel experience and perception of risk and safety, *Journal of Travel Research*, **37** (2), 172- 177.

The Guardian. (2020) 'Outrageous': Qantas criticised for standing down 20,000 workers without pay. Available: https://www.theguardian.com/business/2020/mar/19/coronavirus-qantas-and-jetstar-to-suspend-international-flights-and-stand-down-20000-workers, viewed 10/04/ 2020.

The Telegraph, (2018) Terror Attacks in France: From Toulouse to the Louvre', *The Telegraph,* 24 June, viewed 11/10/2019, https://www.telegraph.co.uk/news/0/terror-attacks-france-toulouse-louvre/

Uhlmann, M. (2016) France's challenges for working out a coherent strategy against violent radicalization and terrorism. A broad (and incomplete) outline', *Sicherheits Politik-Blog*, 20 December, viewed 22/12/2019, https://www.sicherheitspolitik-blog.de/2016/12/20/frances-challenges-for-working-out-a-coherent-strategy-against-violent-radicalization-and-terrorism-a-broad-and-incomplete-outline/

UNESCO. (2019) France, *World Heritage Centre: State Parties*, n.d., viewed 22/12/2019, https://whc.unesco.org/en/stateparties/fr

UNWTO. (2001) *UNWTO Tourism Highlights, 2001 Edition*, UNWTO, viewed 20/10/2019, https://www.e-unwto.org/doi/book/10.18111/9789284406845

UNWTO (2011) *Risk and Crisis Management in Tourism Sector: Recovery From Crisis in the OIC Member Countries.* Available: https://www.sbb.gov.tr/wp content/uploads/2018/11/Ris_and_Crisis_Management_in_Tourism_Sector-.pdf, viewed 21/10/2019.

UNWTO. (2019) *International Tourism Highlights, 2019 Edition*, United Nations World Tourism Organisation, Madrid, viewed 10/11/2019, https://doi.org/10.18111/9789284421152

Walters, G., Mair, J. and Ritchie, B. (2015) Understanding the tourist's response to natural disasters: The case of the 2011 Queensland floods, *Journal of Vacation Marketing*, **21** (1), 101-113.

Weimann, G. and Winn, C. (1994) *The Theater of Terror: Mass Media and International Terrorism.* White Plains, NY: Longman.

White, J. and Gilbert, D. (2020) Australia fires: The numbers that highlight sheer scale of unfolding catastrophe, *The Telegraph,* 2 January, viewed 10/02/2020, https://www.telegraph.co.uk/news/2020/01/02/australian-bushfires-numbers-highlight-sheer-scale-unfolding/

WHO. (2020) WHO Coronavirus Disease (COVID-19) Dashboard, *World Health Organization,* viewed 4/06/2020, https://covid19.who.int/

Young, W.B. and Montgomery, R.J. (1997) Crisis management and its impact on destination marketing: A guide to convention and visitors bureaus, *Journal of Convention and Exhibition Management*, **1** (1), 3-18.

12 Solving Future Problems in the Tourism, Hospitality and Events Sectors

Introduction

The management of tourism, hospitality and events is often consumed with solving problems. Whether it be the daily operation of the business or planning for the future, the manager must make decisions concerning problems that are faced by the organisation. Senior management are often responsible for solving the more complicated problems and issues. These may include having to deal with external stakeholders and the public, who have vested interests in particular outcomes. Solving problems at scale creates challenges for the manager of an organisation or destination. This chapter considers the ways in which problems can be evaluated and solved and alternatives for some of the bigger issues that senior managers and organisations face. It opens with a discussion of traditional approaches to solving problems; those which individuals generally take in their day-to-day lives. It is argued that this does not work well for the types of issues faced by senior managers. Two alternative approaches are introduced which can provide insights into problems and assist in unravelling the issues involved in complex decision-making.

Traditional approaches to problem solving

Solving complex problems is generally more involved than simply following traditional approaches to problem solving, which have been found to be ineffective, particularly around issues of sustainability (Fodness, 2017). However, it is useful to understand this basic approach before considering more involved processes.

The first step in the traditional problem-solving approach is to list the possible alternatives that could solve the problem. Each alternative is then evaluated in terms of its attractiveness and feasibility. Then the decision maker chooses the most attractive option, implements it, and following its implementation evaluates and assesses the outcome against the expected outcome. The solution and assessment can then be used to inform future decision making. Such an approach can be considered suitable for straightforward problems. Consider the situation where someone might be deciding how to get to the airport for their summer holiday. The alternatives could be: ask a friend to drive them; drive themselves; take a taxi or Uber; or take public transport. They would have to consider each option, in terms of time, effort, cost and availability. Once an option is selected, they would implement it and then evaluate it on completion of the task in terms of how it met their needs and expectations.

Use of heuristics

Many problems such as the one above are straightforward and involve breaking the problem down into individual steps and applying rules, or heuristics, to determine alternatives. Rules, for example, guide the decision maker and set standards for consistency in the organisation. Organisations often specify that new purchases require obtaining a specified number of quotes and choosing the lowest price. Heuristics, on the other hand, are a type of rule based on an individual's experience or cultural norms (Nazlan *et al.*, 2018). An example of this decision making might be to purchase an expensive brand based on the premise that better quality lasts longer. Some of the issues that might need to be considered would be obtaining additional information on each alternative. For example, if they drive themselves to the airport, where will they leave their car while they are on holiday? What would be the cost of each alternative? What time would they need to leave to be sure to arrive at the airport in time for the flight? Breaking the problem down in such a manner helps the decision-maker understand the potential advantages and disadvantages of each method. This approach to problem solving has been studied for some time by economists who make the assumption that decision making is rational and seeks to maximise utility. The approach is often referred to as 'linear' because one thing leads to another, and each option can be determined systematically – often with finite outcomes (for example, cost and time) (Taylor, 1947; Mintzberg and Waters, 1985; Collis, and Montgomery, 1995).

Routine problem solving

For business managers, many day-to-day operations can be solved using this straightforward decision-making process. The implementation of many operational tasks in business, and even in our daily lives, is often repetitive and which can become routine: for example, who to purchase from, which airline to use, or what brand to choose. As a result, businesses do not have to reinvent the decision-making process every time and some businesses become specialised in solving problems for other people and businesses. Event management organisations become specialists in running events for others for this very purpose. Other examples include wedding or party planners, or travel agents who solve an assortment of problems associated with their travel related services for their clients.

Managers use these problem-solving techniques every day in both their personal and professional lives, and as a result, the process becomes acceptable practice. The traditional problem-solving approach is also very suited to our fast moving, quick acting economic system where timely decisions and action are preferred over deeper introspection and consideration for consequential implications. Information provided is designed to encourage immediate reaction – for example, when consumers are encouraged to 'buy now' so they don't 'miss out!' Our exchange system also focuses on the tangibility and cost of items discounting other costs and implications. As a result, our economic system favours and benefits a linear approach to problem solving.

Non-linear solutions

However, not all the problems that are faced in the world today are able to be solved in this straightforward linear manner. A range of breakdowns are encountered in our processes that mean the actions of one solution may result in problems for others. For example, if public transport is not operating, or not operating frequently enough, it may not allow people to get to the airport on time or that the waiting time would be too long. If an individual lives in a regional or remote area, some public transport options such as taxis or Uber may not be available. As a result, the individual must drive and leave their car at the airport, resulting in further expenses and an increase in greenhouse gas emissions. Solutions might also be considered out of the reach for some because of the cost involved, or by not being suitably informed of available options and alternatives. Rules may also be too rigid, making tasks difficult to implement.

Wicked problems

'Wicked problems' are complex social or cultural issues which are difficult or impossible to resolve (Rittel and Webber, 1973). Challenges such as solving climate change (environment), obesity (health), terrorism and crisis management (security) and poverty (social justice) are just a few of these complex issues referred to extensively in the literature (see for example,).

While some of these wicked problems have been around for a long time – for example, poverty, famine, inadequate water and education – others are on the rise and new wicked problems are being identified. If these problems are to be solved in the future, alternative approaches to problem solving are necessary. In the next section, two approaches are discussed which may be applied to the tourism, hospitality and event management industries and to business management more generally.

Alternative approaches to solving complex problems

The two approaches introduced within this section are related and can be used together to provide insights and solutions to problems. The first is referred to as 'systems thinking', which has been introduced in Chapter 11, while the second is known as 'scenario planning'. While both approaches are quite different from the standard decision-making processes that have been outlined above, they differ from one another.

One aspect of the traditional approach is the use of historical data and information, which does not take into account current and future changes. This linear approach is often referred to as being 'reductionist', as it attempts to break issues down to the smallest components, which managers can then address individually. Managers also tend to fall into the trap of looking at the symptoms rather than the causes – for example, drought and bushfires rather than the impacts of human-caused climate change. A classic example that many readers may be familiar with is documented in *The Tragedy of the Commons* (Hardin, 1968), which recounts the story of the impact of overgrazing a village common (a communal piece of land in a village). As each farmer adds more and more cattle to the land, the land becomes more degraded and as a result, it cannot maintain the original number of cattle. Subsequently, all the famers and the village suffer the consequences. This simple tale highlights the problem of resource allocation and points to the need for intervention. It also highlights how the failure of the environmental resource was the result of not being adequately managed by the economic system, and suggests that individual decision making cannot be solely relied upon to protect our natural resources.

With this in mind, future managers need to expand their approaches to problem solving by learning about and exploring alternative options to resolve issues. In doing so, they may need to rethink their approach and challenge their assumptions about practices by exploring alternatives for decision-making. Consequently, they will not only be able to benefit their organisation but also that of society and individuals. This is the challenge facing future managers and destination management organisations.

Systems thinking

Sterman (2000: 2) defines systems thinking as 'the ability to see the world as a complex system, to understand how everything is connected to everything else'. As we have seen in Chapter 11 regarding crisis management, systems thinking has been recommended as an appropriate tool to assist tourism operators understand the intricate links between stakeholders who could potentially be involved in dealing with a crisis.

Higgins (2014) points out that systems thinking has emerged from a multi-disciplinary background based on philosophy, sociology, biology, and engineering. The 'hard' sciences, with a focus on quantitative methods, rely on the analysis of data, graphs, loops and flow diagrams to examine and explain the impact of one factor on another, and are grounded in a study of system dynamics (Sterman, 2000). For businesses and business students, system dynamics may have general appeal; however, it is important to note that there is more to understanding systems than just studying the hard sciences. By adopting an approach that draws on both a quantitative and qualitative approaches, researchers and decision makers are able to take a holistic view of the problem. From here the component parts and their interactions may reveal better design solutions and solve some of the larger overarching problems facing the industry.

In developing an understanding of a systems thinking approach, it becomes apparent that the application of traditional problem-solving approaches is not possible in solving some of the bigger issues that face businesses, government and society, such as climate change, unsustainable tourism development and pollution. Another way of explaining systems thinking is in terms of how problem-solvers go about approaching the problem. As mentioned above, looking at the individual components tends to blind the problem-solver to the bigger picture. What becomes important in problem-solving is an understanding the context in which the problems exist (Dutta, 2017).

System design

Systems thinking researchers and planners have adopted the tools of system designers in an attempt to gain an overall perspective of the problems to be solved. This includes the use of two tools – referred to as 'causal mapping' and 'stock and flows' (Sterman, 2000). While it is not possible to provide a detailed account of these tools in this chapter, the provision of a brief outline should help guide the reader who is encouraged to seek more details on the topics from the references located at the end of this chapter.

Causal mapping

Causal mapping is the idea of looking for and mapping the impact of relationships between actors involved in a problem. In looking at any problem, there are multiple actors and their relationships can impact on the actors in a positive (additive) or negative (decreasing) manner. The goal of causal mapping is to identify all the relationships involved in the problem at hand. The approach can be applied to any problem and seeks to answer the question 'what happens if I do this to that?' The previous discussion of problem-solving has proposed how one can implement a solution to a problem through breaking down problems into a series of parts. Here, however, the implications of an activity are being considered and then extended onto other activities involved in the problem. By looking at each of the actions, the researcher can inform the strategic thinking and planning for the issue (Bryson et al., 2004). The following is a simple example to illustrate causal mapping – the problem of eco tourists visiting a popular natural tourist destination.

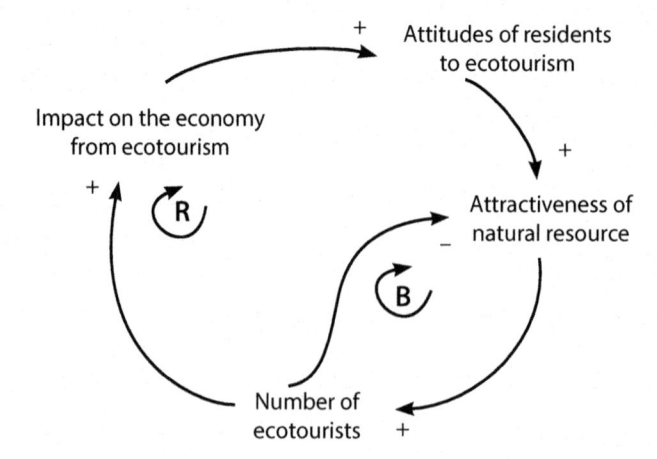

Figure 12.1: Simplified causal map of ecotourism. *Source*: P. Vitartas, 2020.

The starting point is the number of ecotourists located in the bottom of Figure 12.1. Moving up and to the left it can be seen that more ecotourists are positive for the local economy (indicated by the + sign). Likewise, the impact on the local economy has a positive impact on the residents' attitudes toward ecotourism, which in turn has a positive impact on the conservation of the natural resource, which in turn draws more ecotourists to the area. This completes a loop of the system and all the activities have positive signs – this indicates a *reinforcing* loop (evidenced by the R symbol within the circular arrow). This suggests that the process will increase continually until collapse. Within the causal map we can also see another connection from the number of ecotourists to the conservation of natural resources. This time however, as the number of ecotourists increases, there is a detrimental effect (indicated by the – sign) on the conservation of the natural resource. As more ecotourists visit the resource there is impact that degrades the area through overcrowding, waste and pollution. This is referred to as a *balancing* loop (indicated by the B symbol) – as more visit the area then there is congestion that balances the number of ecotourists. If the balance does not occur effectively, the system may still collapse but after a longer period, so intervention of some type may still be necessary.

Developing causal maps involves a considerable amount of research and input from stakeholders. Several meetings between the stakeholders would occur to enable input from each group to be gathered and incorporated into a causal map, as either a positive or negative effect on the issue. Additional research may involve undertaking surveys in the community and among participants to better understand the needs, preferences and attitudes of those involved in the issue.

The above simplified example highlights the potential issues that planners need to consider in mapping for a solution. Options might include limiting the number of ecotourists visiting the area, considerations of infrastructure to support tourism and/or the addition of educational programs to assist in protecting the area targeted to ecotourists and residents.

Stocks and flows

The second tool available to systems thinking planners follows on from the causal mapping process, and allows for the input of quantitative data to measure the impact of the reinforcement and balancing loops indicated in the causal map. In the 'stocks and flows' tool, planners are able to apply quantitative estimates to the steps identified in the causal map and assess the impact of each on the system. The data would be drawn from historical and current data that has been collected for the purpose of the project. Specialised software using the data allows for the simulation of 'what if' scenarios that can be evaluated by the planners.

A popular way of explaining this process is through the use of a bathtub as the representation of a stock. When the tap is turned on, water flows into the bath and is stored. Likewise, when the drain is opened water flows out of the bath. By being able to calculate the inflows and outflows to a system, the level of the bath can be calculated. Causal model elements impact the flows and 'leakages' for a system design. As a practical example, a stock could be the amount of accommodation available in a tourist town. The flow would be the number of tourists visiting the town. As more tourists visit, the stock fills to the point of being full or overflowing. Establishing new or alternative accommodation would see a flow of tourists to this new accommodation, relieving pressure on the existing infrastructure. The approach can be applied to impacts on roads, electricity, waste disposal and any other issues where data is obtained.

Software such as Stella (isee systems, 2020a) and iThink (isee systems, 2020b) are dedicated programs which allow researchers to map causal models and incorporate the stock and flow concept in modelling systems. Simple models can also be developed using spreadsheets.

In the next section, a second approach to systems thinking is outlined – that of scenario planning. In this approach the reader may recognise the connection with systems design as an approach which builds on causal mapping and stocks and flows.

Scenario planning

Today scenario planning builds on the two previously described methods of analysis by considering and testing varying scenarios being considered. Thought to have originated from military planning and intelligence and the use of war games (Brown, 1968), its adoption by businesses emerged in the 1960s (Bradfield *et al.*, 2005). At the time scenario planning was frequently applied to the development of weapons systems, uncertainty around the establishment of the USSR and iron curtain, and the future political system that was emerging. With the advent of computer systems and game theory, a simulation of a greater range of scenarios was possible whilst taking into account data and social interactions. There was also the realisation that much of the current military thinking of the time was not based on 'reasonable expectation' and would result in disastrous outcomes under simulated conditions, particularly around nuclear war. Thus, improved decision-making techniques were welcomed by government planners and business operators. Publications at the time, including *The Year 2000* by Kahn and Wiener (1967) and *Limits to Growth,* from the Club of Rome Reports (Meadows *et al.*, 1972), brought into popularity the idea of using scenarios to provide insights into the future. Some of this work alerted planners to disruptions in the oil industry and environmental conditions for businesses into the future (Wack, 1985).

The key to scenario planning is the development of plans under circumstances of uncertain environments. As previously discussed, traditional linear planning approaches rely on historical data for analysis and may not include realistic data for future events. In those cases, a range of alternatives must be constructed so that they can be tested under differing circumstances. This approach allows managers to understand potential scenarios and consequently anticipate, be more adaptive and better prepared for unforeseen circumstances. In tourism for example, recent research employing scenario planning has considered transport and mobility, demographic and social change, greenhouse gas mitigation, terrorism and crisis management. More generally, there has been interest in understanding global trends in tourism, as well as projecting what future tourists might look like in the period 2030 – 2050 (Yeoman, 2012).

At a local level, planners and tourism operators can benefit from scenario planning by considering the impact of growing tourism on local environments, such as population growth, infrastructure, land and fresh water requirements, natural environment demands, air and water pollution, waste disposal and the overall impact of all of these factors on the attractiveness of a destination. In other cases, the process of undertaking scenario planning has brought together communities and stakeholders, enabling all participants to obtain a much greater understanding of the issues involved and challenges being faced by others. This process provides an opportunity for discussion, but also for insights that can lead to new and novel outcomes and a consensus on potential scenarios.

Process of developing scenarios

The development of scenarios varies considerably; however, the broad steps appear to follow along the lines proposed by Schwartz (1991):

1. Identify the focal issue or decision to be made;
2. Identify stakeholders and basic trends in the local environment;
3. Identify driving forces and key uncertainties affecting the organisation;
4. Rank the driving forces by impact;
5. Select a series of potential scenarios (5-10);
6. Refine and develop the scenarios (3-5);
7. Identify the implications; and,
8. Select leading indicators for evaluation.

Generally, organisations consult specialist teams to assist them in this planning process because of their knowledge and the input they can provide on future trends and industry background. These specialist teams initially work with senior managers to identify and refine management's clarity on what the focal issue(s) is/are and the decisions that need to be made. Internal and environmental scans are then undertaken to identify the key trends and uncertainties that the organisation is facing. At this stage, the scenario developers work with managers and planners in the organisation to work through the driving forces they have identified. This becomes part of the learning process for members of the organisation as they develop an understanding of future issues they will need to consider in their future planning processes. Based on their feedback an initial set of scenarios is developed – this may be as many as ten, but they are then reduced down to an acceptable number – three to five. Once agreed upon, they are then developed into detailed stories or narratives which describe the future organisation or destination, its issues and challenges. These scenarios can then be used by the organisation to help develop thinking and planning among managers for their specific divisions or departments. They can be used to brainstorm implications and potential opportunities that can then be considered in the planning process. Consideration for such plans could lead to a change in direction for the organisation, the development of new product lines or expansion into new markets.

The case study for this chapter presents an example of three scenarios based on a project undertaken by the Business Council of Australia (BCA) in 2004. The project – 'Aspire Australia 2025' was designed to enable Australia to prepare for the future. As can be seen from the scenarios, the reader is challenged to think about what the implications might be for each of the scenarios. As 2025 approaches, the reader might reflect on how well the scenarios have been able to capture our current issues and the business landscape given the development of these scenarios almost 20 years ago.

Frameworks for future decision-making

The traditional approach to decision making, including two alternative approaches, have been discussed in this chapter. There are of course many others that can be considered, each with a focus on a specialist area. Three areas that the reader might also want to consider for further reading include the use of the Delphi method, design thinking and user experience.

☐ **The Delphi Method** (Dalkey and Helmer-Hirschberg, 1962), named after the 'Oracle of Delphi', a Greek priestess known for her prophecies, is predominantly a forecasting method based on anonymous feedback from multiple rounds of consultation with experts in the field of interest. This

results in a group consensus after multiple rounds that is supposedly not biased or influenced by an influential figure or figures in the group.
- **Design thinking** (Arnold, 2016) is not dissimilar to systems thinking and seeks solutions to ill-defined or unknown problems. It is a popular technique used in business and has become popularised through its use by Google and a number of other major technology companies.
- Related to design thinking is the concept of **User Experience Design** or **UX**. Popularised by design engineer Don Norman, students of marketing may have come across the idea of UX. Its focus is on meeting the needs of users, and UX has a customer experience emphasis which may be particularly suitable for the service-related industries (Norman, 2013). Today it is often associated with software or web page design although its conception was in product design. In essence, the aim is to design a product that is easy to use, intuitive and efficient. On occasion, experts may consider it a micro-focused designed strategy, when compared to systems thinking, due mainly to its strong focus on research.

Summary

Decision making for larger problems requires much more than linear thinking. Whether a manager, planner, researcher or other type of decision-maker, approaching problems from a citizen perspective (Tufte, 2017), with consideration for all stakeholders, holds the greatest opportunity for solutions for society. In this chapter the concepts of systems thinking and scenario planning as approaches to thinking about problems were introduced with the aim of findings solutions that include consideration for the many stakeholders that impact on and are impacted by the problem. The use of tools such as causal mapping and stocks and flows encourage examining the problem as a whole and identifying the links from the result of actions. Scenario planning was discussed as it allows managers to not only identify factors that could affect the organisation, but also to share these scenarios to guide the thinking and considerations of people in the organisation. Finally, a number of additional tools managers can consider for use in their decision-making were highlighted, including the Delphi method, design thinking and user experience.

The methods introduced in this chapter cannot be introduced and undertaken by individuals – they need the support and involvement of many members of an organisation, destination or community. The decision maker must also be a communicator and leader who can bring together disparate groups so they can work and be involved in decisions that will impact them and their lives. As Arch Woodside has commented, 'Nurturing integrative thinking is more likely to achieve sustainable tourism and high quality-of-life objectives than is taking an intractable stance' (Woodside, 2009: 214).

Case study: Business Council of Australia scenario summaries from Aspire Australia 2025

The following are extracts from three scenarios presented in *Aspire Australia*, a publication from the Business Council of Australia that uses the scenarios as a planning tool. While the extracts here are not a full recount of the original scenarios, they provide an insight into some of the hypothetical scenarios that are used in Scenario Planning. Readers are referred to the full document listed in the references for the complete scenario details.

Read the case studies and consider how a tourism operator would need to respond to each scenario.

Scenario 1: Riding the Waves

In 'Riding the Waves', governments – both Federal and State – know that further reforms are needed to improve the effectiveness of spending and grow the economy: labour markets, welfare, taxation, superannuation, water rights are all in need of reform. However, a combination of electoral pressure, especially given the rise in single issue interest groups, and a chronic lack of coordination in the political process, brings some to the realisation that Federal and State Governments have not made the necessary hard decisions needed to address the growing problems.

Key areas of progress have included reforming the labour market, reducing tariffs, floating the exchange rate, deregulating the financial sector, reforming corporate governance, taxation reform, developing a national competition policy, reforming financial regulation and introducing new frameworks for monetary and fiscal policy. Inconsistencies between states in the application of their regulatory frameworks, difficulties with competition reform, and lack of coordination between levels of government were all noted as major impediments to the creation of a sufficient infrastructure network for Australia.

As pressures mount, and Australia's rate of economic growth diminishes, governments are left with no choice but to cut spending and increase taxes, even in the face of a dire need to shore up infrastructure and services. Australia's obvious decline, undermining the long-held view that we are the lucky country, galvanises political and community debate away from local issues to national interests.

Scenario 2: Stormy Seas

In 'Stormy Seas', an initial period of high growth gives way to a sustained decline in Asia Pacific stability and security, challenging Australia's international and economic relations. In the early years of this scenario, high Chinese economic growth helps to fuel the Australian economy. China's growth benefits the island states of the South Pacific through generous aid programs, and Beijing becomes the region's largest donor. Canberra worries about stability in Asia and the South Pacific and, in particular, whether this will create a larger illegal immigration flow in future years. Our ties with China and the US overcome worries about the fragility of our nearer neighbours, while our economic strength reinforces the view that Australia's national qualities will protect us from regional shocks.

The decline in Asia Pacific stability weakens the region's economic prospects and erodes Australian markets. However, just as growth in the economy starts to contract, security concerns force Canberra to increase the level of defense and counter-terrorist spending. The region would significantly benefit from stable economic growth, a declining rate of corruption and inefficiency in government and by working hard to avoid military conflicts.

Scenario 3: Changing the Crew

In 'Changing the Crew', a new generation of pioneers creates a sharper-edged Australia, resulting in friction with other generations. Australians are more strongly connected with the rest of the world than ever before, economically and culturally. The new crew include The Armchair crew 65 to 84 years old (2.4 million in 2001); The Cockpit crew 45 to 64 years old (4.2 million in 2001); The Foredeck crew 25 to 44 years old (5.9 million in 2001).

The way of the Foredeck crew is to build up the asset of themselves, and to act swiftly to apply new ideas and technologies to capture projects, niches and fleeting but potentially lucrative market opportunities. The Foredeck crew and the following generation are comfortable with the notion of establishing businesses, running them hard for a period, and then moving on to the next opportunity. The Foredeck crew favours the continuous adjustments and opportunities created by markets, rather than government interventions in economic activity and associated lobbying. The resources sector, once looked down upon by some promoting the advancement of Australia, becomes an unexpected source of wealth for the government and business as the relative price of minerals and energy increases compared to the price of manufactured products. Workforce pressures, particularly the need for skilled professionals, language skills and overseas connections, and the desire for reciprocal opportunities overseas, mean that the Foredeck crew, as they take over, push strongly for increased immigration. The country has had a more successful generational transition than the US, Europe and Japan, in part because of smaller numbers in the Cockpit crew, and because of growth opportunities in the region.

Acknowledgement: The authors would like to thank the Business Council of Australia for allowing the above extracts to be used in this text.

References

APSC. (2018) *Tackling Wicked Problems: A Public Policy Perspective*, Australian Public Service Commission, 12 June, viewed 5/04/2020, https://www.apsc.gov.au/tackling-wicked-problems-public-policy-perspective

Arnold, J. (2016) *Creative Engineering: Promoting Innovation by Thinking Differently*, Stanford CA: John E. Arnold, Jr, viewed 5/04/2020, https://stacks.stanford.edu/file/druid:jb100vs5745/Creative%20Engineering%20-%20John%20E.%20Arnold.pdf

Bradfield, R., Wright, G., Burt, G., Cairns, G. and Van Der Heijden, K. (2005) The origins and evolution of scenario techniques in long range business planning, *Futures*, **37** (8), 795–812.

Brown, S. (1968) Scenarios in systems analysis, in E.S. Quade, W.I. Boucher (eds), *Systems analysis and policy planning: applications in defence*, New York: Elsevier Publishing.

Bryson, J., Ackermann, F., Eden, C. and Finn, C. (2004) *Visible Thinking: Unlocking causal mapping for practical business results*, Chichester: John Wiley & Sons.

Business Council of Australia. (2004) *Aspire Australia 2025: Change is Inevitable, Progress is Not,* 15 March, viewed 5/04/2020, https://www.bca.com.au/scenarios-for-australias-next-20-years

Collis, D.J. and Montgomery, C.A. (1995) Competing on resources: strategy in the 1990s, *Harvard Business Review* **93**, (Jul-Aug), 119-28.

Dalkey, N. and Helmer-Hirschberg, O. (1962) *An Experimental Application of the Delphi Method to the Use of Experts*, Santa Monica, CA: Rand Corporation, viewed 5/04/2020, https://www.rand.org/pubs/research_memoranda/RM727z1.html

Dutta, P. (2017) *Systems Thinking for Effective Managers: The Road Less Travelled*, Los Angeles, CA: Sage Publications.

Fodness, D. (2017) The problematic nature of sustainable tourism: some implications for planners and managers, *Current Issues in Tourism*, **20** (16), 1671-1683.

Hardin, G. (1968) The tragedy of the commons, *Science*, **162** (3859), 1243-48.

Higgins, K. (2014) *Economic Growth and Sustainability: Systems Thinking for a Complex World*, San Diego, CA: Academic Press.

Isee systems (2020a) *Stella Online*, viewed 5/04/2020, https://iseesystems.com/store/products/stella-online.aspx

Isee systems (2020b) *IThink Online*, viewed 5/04/2020, https://iseesystems.com/store/products/ithink.aspx

Kahn, H. and Wiener, A. (1967) *The Year 2000: A Framework for Speculation on the Next Thirty-Three Years*, New York: Macmillan.

Meadows, D.H., Meadows, D.L., Randers, J. and Behrens III, W. (1972) *The Limits to Growth*, Potomac Associates – Universe Books, First Edition (digital version), viewed 11/06/2020, http://collections.dartmouth.edu/teitexts/meadows/diplomatic/meadows_ltg-diplomatic.html

Mintzberg H. and Waters J.A. (1985) Of strategies, deliberate and emergent, in *Strategic Management Journal* **6** (3), 257-272.

Nazlan, N., Tanford, S. and Montgomery, R. (2018) The effect of availability heuristics in online consumer reviews, in *Journal of Consumer Behaviour*, **17** (5), 449-460.

Norman, D. (2013) *The Design of Everyday Things,* New York: Basic Books.

Rittel, H. and Webber, M. (1973). Dilemmas in a general theory of planning. *Policy Sciences*, **4** (2): 155–169. doi:10.1007/bf01405730.

Schwartz, P. (1991) *The Art of Long View: Planning for the future in an uncertain world,* Wiley: First edition , New York: Doubleday.

Sterman, J. (2000) *Business Dynamics*, New York: McGraw-Hill.

Taylor, F. W. (1947) *The Principles of Scientific Management,* New York: W.W. Norton.

Tufte, T. (2017) *Communication and Social Change: a Citizen Perspective*, New York: Wiley.

Wack, P. (1985) Scenarios: Uncharted waters ahead, *Harvard Business Review* **63** (5), 72-89.

Woodside, A. (2009) Applying systems thinking to sustainable golf tourism, *Journal of Travel Research,* **48** (2), 205-215.

Yeoman, I. (2012) *2050 – Tomorrow's Tourist*, London: Routledge.

13 The Demise of Tourism?

Introduction

This chapter investigates the possibility that the tourism industry, as we currently know it, will significantly change and perhaps will not exist in the future. Some topics will be discussed in the realm of plausible futures, meaning that they may not happen, however it's a possibility and in the event it does occur, the tourism industry should be prepared. The topics discussed in this chapter include having an understanding of the reliance of natural resources in the travel, hospitality and event sectors; global population growth; food security; the impact of war on tourism; and the moral considerations associated with certain tourist experiences. Pandemics including Covid-19 are mentioned in Chapters 2, 4, 6, 7, 9, 11 and 14. The case study focuses on food security and the dangers of depleting the quantity of food around the globe, along with the availability of quality nutritional food. It also explores the changes in the types of food supplied to the tourism, hospitality and event sectors, and provokes consideration of the disparity between the wealthy and populations born into poverty.

Plausible futures

Plausible futures fit within the realm of future studies that is considered an under-researched area in tourism (Frost *et al.,* 2014). Authors such as Amelung and Viner (2006), Walton (2008), Bergman *et al.* (2010), Yeoman (2013) and Postma *et al.* (2017) have all undertaken research in this area. Walton (2008) suggested a plausible future is the possibility of something occurring but it may not happen. Strickland (2012) and Yeoman (2012) described it as a plurality of futures which is not necessarily a predicted trend, making it an ideal structure for discussing global events and concepts of the future. This approach gives structure to the discussion and allows the authors the ability to explore ideas that seem possible, however, they may be rejected in the future. The reasons for not coming to fruition may be due to cost, changing technologies, absolute need to find other solutions, best practises, public opinion, government intervention and environmental concerns. Although no solutions are deliberately given here, the direction of future research is suggested for scenario planning as discussed in Chapter 12.

Pandemics including Covid-19 are mentioned though out this book due to its devastating impacts on THE, international travel, global economies, health systems and mortality rates.

Natural resources in the THE sectors

When evaluating natural resources required for the tourism, hospitality and events sectors, studies have been heavily reliant on demand and supply factors in terms of visitors, commercial businesses, governments and non-government entities (Clawson and Knetsch, 1963). For instance, in evaluating nature-based tourism, Priskin (2001) suggested 'physical recreation characteristics, level of development, management, intensity of use and anticipated behavioural classes' would usually be evaluated (Priskin, 2001: 640). From this research, a further four categories regarding natural resources have been identified: 'attraction diversity; accessibility; supporting infrastructure and the level of environmental degradation' (Priskin, 2001: 642). With the current debate regarding climate change, it is useful to have a discussion focussing on the actual natural resources required by today's tourism sector and what resources may be required in the future.

There are examples of natural tourism based on sand and sea; however, natural resources are generally categorised as minerals, fuels and agricultural resources (Acar, 2017). To give perspective of the sheer number of available natural resources, the Mineralogical Society of America (2019) suggested there are over 3800 minerals identified worldwide with between 30-50 additional minerals added annually. To capitalise on these riches, Acar (2017: 2) argued that some countries have benefited economically by taking advantage of these resources such as Australia, Canada, Kuwait, United Arab Emirates, Norway and Botswana for oil, gas and coal. However, these mining industries do not overtly affect tourism sectors although there is evidence that this is changing. The following paragraphs investigate in more detail.

Minerals

Minerals do not tend to affect the tourism sectors unless an attraction is created. For example, Sovereign Hill in Ballarat is 'Australia's foremost outdoor museum. Sovereign Hill re-creates Ballarat's first ten years after the discovery of gold in 1851, when thousands of international adventurers rushed to the Australian goldfields in search of fortune' (Sovereign Hill, 2020). Not only is Sovereign Hill a major attraction, tourists are still able to pan for gold (a natural resource). In South Africa, the Cango Caves have been a popular tourist being 'situated in a limestone ridge parallel to the well-known Swartberg Mountains, you will find the finest dripstone caverns, with their vast halls and towering formations' (South Cape Tourism, 2020). This is a way to exploit natural resources in a controlled way. Museums located world-wide may also display collections of precious gemstones and/or mineral rocks for tourists to view. However, it appears, unless the natural resources are adapted into an attraction, they tend not to be of interest to the tourism industry. Note, construction of facilities using natural resources such as cement is not considered in this context although they may form part of the attraction.

Fuels

Fuels, on the other hand, affect the tourism industry immensely. Gossling (2000) suggests fossil fuels used in the tourism sector can be identified within two categories; energy consumption needed for travel, and energy consumption within a destination.

Energy consumption for travel refers to fuels required for all types of travel options including airplanes, trains, cars, buses, jet skis, scooters, motorcycles, hydrofoils, and other mechanical transport options. It does not include fuel for living creatures such as food for horses or humans riding a pushbike.

Energy consumption within a destination includes operation of tourist facilities such as hotels, attractions and restaurants. Within these facilities, electricity, water and climate control for example all need natural resources to power them.

Currently, the main forms of fuels are sourced from coal, natural gas, hydro, nuclear power, wind power, solar power and oil. There are other fuel alternatives including ethanol, hydrogen, propane, biodiesel, methanal, P-series fuels (blend of ethanol, natural gas liquids and methyl tetrahydro-furan), wave power, geothermal, radiant power and biomass fuels (Carrette et al., 2000). Only some of these fuels are 100% renewable and it is becoming more socially conscious to transfer to renewable energies for future generations. In response, some governments are trying to balance the cost, source of fuel and environmental implications (climate change) regarding the type of fuel available. There is no global consensus on the preferred fuel source, however many people in the global community would prefer to invest in fuels that have no carbon emissions and do not contribute to climate change. It is plausible that all countries adopt renewable energy in the future although the expectation that it will occur quickly is low, even though alternative energy supplies are inevitable (Cheng, 2017).

Agricultural resources

Agricultural resources consist 'of activities which foster biological processes involving growth and reproduction to provide resources of value. Typically, the resources provided are plants and animals used for fibre and food, although agricultural products are utilised for many other purposes' (Lehman *et al.*, 1993: 127). Additionally, the word 'sustainable' often appears prior to the word agriculture implying that the resources will occur more than once, such as more than one crop of wheat or more than a generation of raising animals.

In response to a growing demand for food and food products for the health conscious, organic agriculture is becoming a phenomenon with new farming techniques that eliminate the need for using foreign interventions such as hormones or pesticides (Willer *et al.*, 2018). In terms of the hospitality sector, offering alternatives such as sustainably sourced, organic or free-range food is a growing trend in both developing and developed countries. This stems from the dissemination of agriculture knowledge through education programs and its application on farms (Quesada-Pineda *et al.*, 2018). It is plausible that all food produced for human consumption in the future will be ethically and organically sourced with minimal impact on the environment. Young and McCoy (2016: 43) argued 'across the board, consumers are indicating greater interest in issues of healthier eating, ethical sources, organic farming, gluten free, and company ethics'. The agricultural economy is slowly responding to the demands of the global consumers and changing their practices where possible.

However, this is not always the case. As discussed in Chapter 12, in the 'tragedy of the commons', overuse of natural resources can result in the loss of viable land. For example, Walker and Meyers (2004) highlighted the detrimental effects of agricultural overharvesting in the USA, causing natural salt levels to rise to the surface from hidden water tables below, only to destroy the natural bio and eco-systems on the surface that cannot easily (if at all) be reversed. Kroon *et al.* (2016) highlighted how land-based run-off pollution is causing irreversible damage that is detrimental to the resilience and health of the Great Barrier Reef in Australia, in addition to reef bleaching through climate change. In Europe, some agricultural practices appear to be disproportionate by economic factors, leading to poor agricultural decisions and depletion in harvest capacity or abandonment of land (Strijker, 2005). In Sumatra, Indonesia, 'millions of hectares of pristine tropical rainforest were destroyed in 2018, according to satellite analysis, with beef, chocolate and palm oil among the main causes' argued Carrington (2019: 1). Africa and South America also have their own but similar agricultural challenges.

Instances such as these often place financial benefit ahead of ethical or environmental concerns. However, it is not all doom and gloom for the tourism industry. Consumer attitudes are changing, and all industries will need to address their concerns. An ideal place to begin is with the morality and culture of individual businesses that flows on to sector wide practices which are changing for the better.

Morality in tourism

Evidence from academic literature suggests that morality in tourism exists; however, Mostafanezhad and Hannam (2016) argued this topic is greatly under-researched. Castaneda (2012) suggested that previous studies highlighting morality are based on cause and effect principles and morality is inherently negative when applied to most tourism products. He also argued that there are many more considerations regarding morality that affect diverse stakeholders, governments, non-government agencies, communities, policy makers and other businesses. The current approach towards using the term morality as being either 'good or bad' has shifted to ethical concerns, as it encompasses more than just investigating the impacts on the local communities (Castaneda, 2012: 48).

There have been studies which have highlighted the moral and ethical concerns from the supply and demand point of view, and have considered how these concerns impact the local environment in which the tourism experience is offered (Wilk, 2001; Butcher, 2003; Duffy and Smith, 2004). Opinions regarding moral and ethical concerns can vary depending on the reliance of the tourist experience. For instance, a study published by Haralambopoulos and Pizam (1996: 503) investigated:

> the impacts of tourism, as perceived by the residents of Pythagorion, a well-established tourism destination on the Greek island of Samos. Interviews conducted with heads of households revealed that residents not only supported the current magnitude of the tourism industry but also favoured its expansion. Despite this, the respondents identified a number of negative tourism impacts, which, in their opinion, affected the town. These impacts included high prices, drug addiction, vandalism, brawls, sexual harassment and crimes. The study reconfirmed that those respondents who were economically dependent on tourism had more positive attitudes towards the industry than those who were not dependent on it.

This research highlighted financial sustainability as an overarching theme as to the continuation of offering tourism experiences, even though significant negative impacts were identified by the local community. This study did not reveal if the local community was trying to reduce the negative impacts that may morally appease the other populations who are not directly reliant on the financial rewards tourism can provide. Conversely, there is evidence that tourism experiences can be offered that negate the adverse effects of tourism. The Kingdom of Bhutan in the Himalayas for example, sets a limit of 100,000 tourist visas per year based on a tariff system. International visitors must pay a minimum of US$250 per day which includes three star-rated hotel accommodation, three Bhutanese meals daily, a driver, a guide and entry into all visitor attractions. This unique tariff system for issuing visas is strictly controlled by the Bhutanese government to ensure international visitors spend a certain amount of money and the government can predict the revenue generated by tourists (Strickland, 2019). It also spreads the wealth through the year in different geographical locations and supports the sustainability principles of the country. The Tourism Council of Bhutan's (2019) website states:

> Bhutan's tourism sector is regarded as one of the most exclusive travel destinations in the world. Today tourism is a vibrant business with a high potential for growth and further development. The Royal Government of Bhutan adheres strongly to a policy of high value, low impact/volume tourism which serves the purpose of creating an image of exclusivity and high-yield for Bhutan.
>
> **Vision:** To promote Bhutan as an exclusive travel destination based on Gross National Happiness (GNH) Values.

Opening up to international tourism from the 1980s, the Kingdom of Bhutan has been able to witness both the positive and negative impacts of tourism and strives to have a realistic balance between tourism and environmental, cultural and spiritual awareness through the principles of Gross National Happiness (GNH). Although a unique concept originating from Bhutan, GNH is currently studied by other organisations as guiding principles to aspire to and a model for the mind, body and soul.

Studies have suggested that tourism experiences only for human pleasure are not something that should be pursued (King et al., 1993). Other moral and ethical concerns pertaining to tourism requires consideration regarding the impacts on the local communities, tourism stakeholders, the environment and the economic value added to geographical location. Culture, cuisine, heritage, language, textiles and architecture also need deliberation before a tourism experience is to be introduced otherwise the demise of tourism may, in fact, be the tourism product and service itself through loss of authenticity and the tourist experience. Figure 13.1 depicts religion, architecture and western visitors in Bhutan.

Figure 13.1: Western tourists in Bhutan climbing to Tiger's Nest in the distance. *Source:* P. Strickland, 2020.

Impact of war on tourism

War and tourism have been interrelated for centuries. Historians often recount the occurrence of war at historical locations, as it may be of interest to tourists and the general population alike. Butler and Suntikul (2013) suggested that some war sites can positively benefit from tourism whereas others do not. These authors highlighted the 'complex and dynamic set of relationships between tourism and war' that include political, economic, psychological and ideological elements that take into consideration the location, history, political stability, transportation and economic value of creating tourism experiences pre, during and post war (Butler and Suntikul, 2013: 2). Evaluating war sites and potential tourism attractions in this way assists in determining the creation of a tourism experience based on negativity and darkness of the occurrence.

There are many examples of how war has had a dramatic impact on tourism experiences around the globe. Conflict has always been present in many forms. Vietnam is an example of a country that has re-invented wartime and conflict sites as a tourism experience that entice comparatively large tourist visitation. There are many locations that have been war afflicted. Some of the main sites popular for international tourists include the Cu Chi tunnels, Reunification Place, Ap Bac Battlefield, Ben Het Special Forces Camp, the Demilitarised

Zone in Quang Tri province, Hue Citadel, Vinh Long Army Airfield and Long Tan battlefield. These all involve conflicts at individual locations and have a rich and well-documented history of events. Other wartime experiences include manmade structures such as the Vietnam War Museum, My Son temple precinct and the Hoa Lo Prison. These sites reflect the effects of war, however, may not have formed part of any major battle although still contribute to Vietnam's wartime heritage.

Regarding tourism and war heritage, Henderson (2000: 269) suggested 'the dilemma is one of achieving a satisfactory balance between education and entertainment while providing an appropriate experience for visitors who come with different needs and expectations. Political circumstances represent an added complication'. Conversely, Alneng (2002: 461) argued that the commoditisation of Vietnamese war heritage is largely based on 'distinct phantasms from western tourists that has been cemented into pop colonial fantasy' that creates the illusion that war sites are a must-see place to visit. Although there are differences of opinion on how to promote, respect and achieve economic benefits of war-torn sites, there will always be a desire to learn, educate and respect the past of the fallen. Most tourists would agree that we must learn from the past and not let war occur again, which is plausible but not yet a reality. Changes in attitudes towards the perceived enemies (opposing sides of war) also needs to occur for all concerned to move forward and have successful tourism experiences. What is evident, if war breaks out, tourism almost completely stops. Governments issue travel warnings, travellers cannot obtain travel insurance and therefore, generally stay away. From previous conflicts, war instigates the demise of tourism in that region.

Impact of global population growth on tourism

In 2019, the world's population was 7.7 billion with an estimated 28% global population increase by 2050 (Population Reference Bureau, 2020). Hence, global population growth is a concern to many (Ehrlich and Ehrlich, 1990). The two main arguments put forward is that in order to increase the standard of living and life expectancy, the population must increase and have more people contributing to the economy and paying taxes. Alternatively, there are large numbers of the population who believe that by over-populating the planet, we are depleting our natural resources at a rapid rate in which, resources will eventually run out. In either case, it is up to humans to come up with solutions (Vorosmarty et al., 2000). For example, airplanes are becoming larger, more efficient, carrying more passengers, yet using less fuel (Kokjohn et al., 2011). Many airlines have carbon neutral programs in which trees are planted to offset carbon emissions. As air travel is becoming more affordable and with a growing middle class in regions such as South-East Asia that can now afford long distance travel, the need for regulation and government control has never been greater (Li *et al.*, 2014).

Similarly, the cruise ship industry is increasing with more and more passengers opting for a cruise as a preferred holiday choice (*Jones et al.*, 2019). In response, cruise ships are becoming self-sufficient, through water purification, waste disposable, recycling, power generation and low environmental impacts. This growth in passenger numbers and the social conscious of passengers has pushed the cruise ship industry to find solutions that are tolerable to tourists. In fact, many passengers believe that their impact on the environment by taking a cruise is so low that it is a conscionable way to travel (Jones *et al.*, 2019).

What about the countries that have large populations and are now able to afford to travel and undertake tourist activities for leisure? India for example, has a rapidly growing middle class and more people are opting to travel (Kharas, 2017). How does the industry respond to these growing demands? First, transport must be addressed. Without adequate transport options, the local population cannot move. This means more airplanes, more trains, more cars and more buses. Second, it must be affordable. Creating transport options that people cannot afford is not sustainable in the long-term. Third, access to the chosen destination should be considered. This is generally controlled by visas, tariffs and general tourist numbers. Machu Picchu in Peru for instance originally only allowed 350 visitors per day in the 1980s. Due to popularity, in 2018, tourist numbers swelled to 4300 tourists per day, according to the local newspaper the Peru Telegraph (2019). Thirty years ago, the tourist policy in Machu Picchu was to preserve the natural and historic significance of the site by controlling visitor numbers as to not over-crowd an already popular destination. Obviously, this policy has changed over the decades due to the revenues tourists can generate. However, not only historical sites worldwide are overwhelmed with tourists. Cities such as Barcelona in Spain have allowed vast tourist numbers with no restrictions, leading to negative impacts for the local people and annual visitors hitting seven million plus (Blanco-Romero *et al.*, 2018).

Many countries also base the success of tourism on the number of visitors and the average spend during their stay. If this is going to be the evaluation benchmark for tourism success, then an increase in populations that travel should be encouraged. Otherwise, what is the alternative? Only the rich can travel? Only people with high community status? Only people from certain countries? Solutions can be found but who will set the precedent? Overtourism is currently being debated with the Covid-19 pandemic and many tourism academics are suggesting that 2020 is the time for change. Furthermore, population growth will not instigate the demise of tourism but more than likely contribute to its growth.

Food security

Global food security has been a concern for some time (Wheeler and Von Braun, 2013). Although it can be argued that the world has enough food to feed the current human population, it is not guaranteed in the future. Today's agricultural methods are generally sustainable but also rely on the natural environment which mankind cannot control, such as hurricanes, floods, drought, earthquakes and other natural events can destroy crops and livestock (Leaning and Guha-Sapir, 2013). In response, humans generally adapt to these conditions. When food stocks are low, food prices go up. When food is plentiful, food prices go down. Some governments such as France heavily mitigate these natural conditions for farmers through subsidies to try to keep prices consistent (Drewnowski and Darmon, 2005). Other governments let the forces of a free trade market dictate the cost of food and eventually the market will correct itself such as in Australia and New Zealand (Worsley and Scott, 2000). Either system supports extensive farming that has been harvested this way for hundreds, if not thousands of years. However, in the current global environment, is this way of agriculture sustainable in the long-term? With a growing population, climate change, disparity in wealth and the need for nutritious fresh produce, what will occur in the future?

Most people would concur that depleting natural resources to extinction is not sustainable long-term (Auty, 2007). So why do some humans keep partaking in these practices? The most logical argument is human survival and to support the immediate family, the village or community (Anwar et al., 2014). In all species, the natural instinct is to do whatever it takes to survive. This is why forests are cut down, open cut mines are created and the oceans overfished. What is the alternative? A few options have been put forward for consideration. First, people can start growing their own crops on their own properties. Any piece of vacant land can be cultivated into a micro-garden subsidising the amount of household vegetables. This often leads to trade with others when crops are plentiful and to exchange different produce. World War Two (WWII) saw this occur when food supply was low due to supplying the soldiers in countries such as Russia, England, France and Australia. Second, the creation of roof top gardens can occur (refer to Figure 13.2). Singapore has many examples of how gardens may be introduced into the city roof top locations (Astee and Kishnani, 2010). This not only allows for the creation of more vegetable and salad produce, but also reduces the effects of carbon emissions in the city and thermal heating conditions of the building. Third, high rise buildings can become city farms. As each level of a hotel can be built, so can high rise agriculture farms. This is not only creating inner city agricultural farms, it is bringing fresh produce closer to consumption and reduces the need for extensive transport. Additionally, there are examples of combating the natural elements by creating artificial rays that mimic the sun. In a town called Port Augusta in rural South Australia there is a large solar tower that reflects sunlight down to grow tomatoes at a rapid rate. The yield is three times greater than the use of natural sunlight. Consequently, more crops can be grown in less time and at a lower cost with a greater yield leading to increased revenues. Furthermore, a company called Apeel is researching the life expectancy of fruits and vegetables. The CEO and founder James Rogers stated 'we look to nature to help the global food system-and all who depend on it-improve quality, reduce waste and ensure an abundant future for our planet' (see additional resources). This means by increasing the natural cover (skin) of some produce, the food should last longer and not go to land fill. Estimations by the National Resource Defence Council (2020) in the USA of edible food that goes to land fill is between 40-60%. In other countries such as Zimbabwe it is much lower, due to the disparity of wealth and significantly greater number of poorer communities (Chep, 2020).

Figure 13.2: Roof Top Garden in Melbourne, Australia. *Source:* P. Strickland, 2020.

So how does food security relate to the tourism and hospitality industries? In Chapter 4, a discussion was presented regarding the importance of food to humans and the industry. Every tourist needs to eat, drink and sleep. Nonetheless, how much do we actually need to eat? Nutritionists suggest is can be as low as 5000kj per day. The recommended kilojoule intake is currently 8700 in Australia. In Brazil the daily intake is 13,050 due to their comparatively active lifestyle. If food shortages occur in the future, who gets to eat? The wealthy? The ill and infirm? The elderly? The tourist? In many hotels, resorts and cruise ships, buffets are extremely popular. 'All you can eat' for a set price. Is this the way we should be presenting food if food shortages occur ? If this is the case, will only tourists be able to dine on an abundance of food? Cuba is an example of the disparity between the seemingly wealthy international tourist and poorer locals. Many large resorts are attractive to Canadian and European tourists. All the resorts have buffets with a few exclusive a la carte restaurants. The buffet is open for breakfast, lunch and dinner. Tourists can consume as much as they like during these times. Other food is available 24 hours a day, such as pizza and hot dogs. Conversely, in the early 2000s the resort staff were queuing for food rations in their local towns and villages, as there was not enough food to feed all the people. This was due to many factors although largely the result of sanctions imposed on the communist government by other nations. Consequently, Cuba had to become self-sufficient and grow their own produce and livestock. As the sanctions restricted chemical imports, Cuba became an organic farming country and now is a world leader in organic farming techniques, teaching other countries on food sustainability (Altieri and Funes-Monzote, 2012).

The bigger question is regarding the quality of the food. Without nutritional content, life is potentially not sustainable. Therefore, all foods produced and supplied should be nutritious and healthy for human consumption. This means foods supplied should contain all the proteins, enzymes, amino acids and vitamins that humans require. It is plausible that foods produced and provided should be organic and nutritious. Perhaps this will lead to the reduction of highly processed convenience foods and restaurants.

The disparity between the wealthy and the poor is increasing (Fadulu, 2019). The wealthy can afford highly nutritious foods yet often choose taste over nutritional value. Also, people born into a comparatively poor environment may not have this option and opt for foods that are less nutritious, such as hamburgers and chips due to the cost. In contrast, in continents such as Africa, the poor often have a highly nutritious and organic diet due to self-sufficiency and growing their own produce. In the future, the proposition is that the wealthy have choices whereas the poor do not have as many options.

In terms of tourism, food wastage is a major concern and buffets are often part of the problem. Should restaurants reduce the number of food buffets in an attempt to reduce food wastage? Should restaurants supply more nutritious foods? Should nutritious food only be available to those who can afford it? It is plausible that all humans can have nutritious food available however change is required to make organic, healthy and nutritionally based products more readily available. This will only happen if change occurs through customer expectations, supplier adaptations, public health information and perhaps government intervention as a last resort. Future research should concentrate on meeting the expectations of customers, food trends, following recommended government guidelines but also affordable and nutritionally dense foods rather than convenience foods that have little to no nutritional value. Food wastage will not be the demise of tourism, however it is predicted that food wastage will be on the radar in the future.

Summary

In conclusion, the demise of tourism is a myth. Through plausible futures methodology, researchers can give probable scenario planning in which to investigate global issues. In this chapter, natural resources, tourism morality, impacts of war, increasing global population growth and food security were highlighted as specific illustrations of how tourism can be affected. Although tourism experiences will change depending on circumstance, it does not mean tourism will fail or even decline. The tourism industry simply needs to adapt to the changing conditions and ideally take into consideration the overall economic, environmental, technological, cultural, political and nutritional concerns to offer tourism experiences to the benefit of all, not just a select few. There is only one certainty: that nothing is certain.

Case study: Food security

View the clip https://www.youtube.com/watch?v=VCYeLuURxRM

Discussion questions

1 The video clip suggests that there is currently enough food to feed at least 11 billion people: however, an inadequacy in food distribution is the main issue.

   a) Should paying tourists receive better quality and nutritional food than a person from a lower socio-economic background?

   b) Should paying tourists receive food in abundance, such as eating from buffets?

2 How can the tourism, hospitality and event sectors actively change to reduce food wastage and perhaps re-direct nutritional food options to people born into poverty?

Additional resources

Apeel Fruit and Vegetables: https://apeelsciences.com/our-story

Chep Food Reduction and Wastage: https://www.chep.com/zw/en/consumer-goods/solutions/corporate-social-responsibility/food-waste-reduction

City Farms: https://popupcity.net/top-5-of-the-greatest-urban-rooftop-farms/

Gross National Happiness: https://www.youtube.com/watch?v=3rNGzD5fWAo

Port August Tomato Farm: https://reneweconomy.com.au/world-first-solar-tower-powered-tomato-farm-opens-port-augusta-41643

Singapore Roof Top Gardens: https://frenzeelo.blogspot.com/2012/05/6-best-roof-top-gardens-to-visit-in.html

References

Acar, S. (2017) *The Curse of Natural Resources: A developmental analysis in a comparative context.* 6th edn, New York: Palgrave-Macmillan.

Alneng, V. (2002) 'What the fuck is a Vietnam?' Touristic phantasms and the popcolonization of (the) Vietnam (War)', *Critique of Anthropology,* **22** (4), 461-489.

Altieri, M.A. and Funes-Monzote, F.R. (2012) The paradox of Cuban agriculture, *Monthly Review,* **63** (8), 23-33.

Amelung, B. and Viner, D. (2006) Mediterranean tourism: exploring the future with the tourism climatic index, *Journal of Sustainable Tourism,* **14** (4), 349-366.

Anwar, R., Sitorus, S.R.P. and Fauzi, A.M. (2014) Achievement of Indonesian sustainable palm oil standards of palm oil plantation management in east Borneo Indonesia, *Journal of Biology, Agriculture and Healthcare*, **4** (9), 1-5.

Astee, L.Y. and Kishnani, N.T. (2010) Building integrated agriculture: Utilising rooftops for sustainable food crop cultivation in Singapore, *Journal of Green Building*, **5** (2), 105-113.

Auty, R.M. (2007) Natural resources, capital accumulation and the resource curse, *Ecological Economics*, **61** (4), 627-634.

Bergman, A. Karlsson, J. and Axelsson, J. (2010) Truth claims and expectancy claims: An ontological typology of future studies, *Futures* **42** (8), 857-865.

Blanco-Romero, A., Blázquez-Salom, M. and Cànoves, G. (2018) Barcelona, housing rent bubble in a tourist city. Social responses and local policies, *Sustainability*, **10** (6), 2043, 1-18.

Butcher, J. (2003) *The Moralisation of Tourism: Sun. Sand... and Saving the World*. New York: Routledge.

Butler, R. and Suntikul, W. (2013) Tourism and war. An ill wind?, in R. Butler and W. Suntikul (eds.), *Tourism and War*, London: Routledge, pp.1-11.

Carrette, L., Friedrich, K.A, and Stimming, U. (2000) Fuel cells: principles, types, fuels, and applications, *ChemPhysChem*, **1** (4), 162-193.

Carrington, D. (2019) Death by a thousand cuts: vast expanse of rainforest lost in 2018, *The Guardian*, 25 April, viewed 11/06/2019, www.theguardian.com/environment/2019/apr/25/death-by-a-thousand-cuts-vast-expanse-rainforest-lost-in-2018

Castaneda, Q. (2012) The neoliberal imperative of tourism: Rights and legitimization in the Unwto global code of ethics for tourism, *Practicing Anthropology*, **34** (3), 47-51.

Cheng, J. (2017) 'Introduction', in J. Cheng (ed.), *Biomass to Renewable Energy Processes*, 2nd edn, New York: CRC Press, pp. 1-7.

Chep. (2020) *Food Waste Reduction*, n.d., viewed 23/01/2020, www.chep.com/zw/en/consumer-goods/solutions/corporate-social-responsibility/food-waste-reduction

Clawson, M. and Knetsch, J.L. (1963) Outdoor recreation research: Some concepts and suggested areas of study, *Natural Resources Journal*, **3** (2), 250-275.

Drewnowski, A. and Darmon, N. (2005) Food choices and diet costs: an economic analysis, *The Journal of Nutrition*, **135** (4), 900-904.

Duffy, R. and Smith, M. (2004) *The Ethics of Tourism Development*, London: Routledge.

Ehrlich, P.R. and Ehrlich, A.H. (1990) *The Population Explosion*, London: Hutchinson.

Fadulu, L. (2019) Study shows income gap between rich and poor keeps growing, with deadly effects, in *New York Times*, 10 September, viewed 22/01/2020, https://www.nytimes.com/2019/09/10/us/politics/gao-income-gap-rich-poor.html

Frost, W., Laing, J. and Beeton, S. (2014) The future of nature-based tourism in the Asia-Pacific region, *Journal of Travel Research*, **53** (6), 21-732.

Gossling, S. (2000) Sustainable tourism development in developing countries: Some aspects of energy use, *Journal of Sustainable Tourism,* **8** (5), 410-425.

Haralambopoulos, N. and Pizam, A. (1996) Perceived impacts of tourism: The case of Samos. *Annals of Tourism Research,* **23** (3), 503-526.

Henderson, J. C. (2000) War as a tourist attraction: the case of Vietnam, *International Journal of Tourism Research,* **2** (4), 269-280.

Jones, P., Comfort, D. and Hillier, D. (2019) Sustainability and the world's leading ocean cruising companies, *Journal of Public Affairs,* **19** (1), e1609, 1-10.

Kharas, H. (2017) *The Unprecedented Expansion of the Global Middle Class: An update,* Washington DC: Brookings.

King, B., Pizam, A. and Milman, A. (1993) Social impacts of tourism: Host perceptions, *Annals of Tourism Research,* **20** (4), 650-665.

Kokjohn, S.L., Hanson, R.M., Splitter, D.A. and Reitz, R.D. (2011) Fuel reactivity controlled compression ignition (RCCI): a pathway to controlled high-efficiency clean combustion, *International Journal of Engine Research,* **12** (3), 209-226.

Kroon, F.J., Thorburn, P., Schaffelke, B. and Whitten, S. (2016) Towards protecting the Great Barrier Reef from land-based pollution, *Global Change Biology,* **22** (6), 1985-2002.

Leaning, J. and Guha-Sapir, D. (2013) Natural disasters, armed conflict, and public health, *New England Journal of Medicine,* **369** (19), 1836-1842.

Lehman, H., Clark, E.A. and Weise, S.F. (1993) Clarifying the definition of sustainable agriculture, *Journal of Agricultural and Environmental Ethics,* **6** (2), 127-143.

Li, J., Granados, N. and Netessine, S. (2014) Are consumers strategic? Structural estimation from the air-travel industry, *Management Science,* **60** (9), 2114-2137.

Mineralogical Society of America. (2020) *Mineralogical Society of America,* n.d., viewed 10/01/2020, http://minsocam.org

Mostafanezhad, M. and Hannam, K. (2016) *Moral Encounters in Tourism,* Abingdon, Oxon: Routledge.

National Recourse Defence Council. (2020) viewed 2/02/2020, https://www.nrdc.org/

Peru Telegraph. (2019) How many tourists visited Machu Picchu in 2018?, 19 March, viewed 5/02/2020, https://www.perutelegraph.com/news/peru-travel/how-many-tourists-visited-machu-picchu-in-2018

Population Reference Bureau. (2020) *World Population Data,* viewed 4/02/2020, https://www.prb.org/worldpopdata/

Postma, A., Cavagnaro, E. and Spruyt, E. (2017) Sustainable tourism 2040, *Journal of Tourism Futures,* **3** (1), 13-22.

Priskin, J. (2001) Assessment of natural resources for nature-based tourism: the case of the Central Coast Region of Western Australia, *Tourism Management,* **22** (6), 637-648.

Quesada-Pineda, H., Smith, R. and Berger, G. (2018). Drivers and barriers of cross-laminated timber (CLT) production and commercialisation: A case study of western Europe's CLT industry. *BioProducts Business*, 29-38.

South Cape Tourism. (2020) 'Developing tourism in the Southern Cape', viewed 14/07/2020, http://www.southcapetourism.com/wpbdm-directory/cango-caves/

Sovereign Hill. (2020) About Sovereign Hill, viewed 7/02/2020, https://sovereignhill.com.au/about-sovereign-hill

Strickland, P. (2019) Hotel and tourism management education in Bhutan: The creation of the Royal Institute of Tourism and Hospitality, in C. Liu and H. Schanze (eds), *Tourism Education and Asia*, Singapore: Springer, pp. 107-124.

Strijker, D. (2005) Marginal lands in Europe—causes of decline, *Basic and Applied Ecology*, **6** (2), 99-106.

Tourism Council of Bhutan. (2019) *About TCB*, viewed 5/02/2020, https://www.tourism.gov.bt/about-us/about-tcb

Vorosmarty, C.J., Green, P., Salisbury, J. and Lammers, R.B. (2000) Global water resources: Vulnerability from climate change and population growth, *Science*, **289** (5477), 284-288.

Walker, B. and Meyers, J. (2004) Thresholds in ecological and social–ecological systems: a developing database, *Ecology and Society*, **9** (2), viewed 10/06/2020, https://www.ecologyandsociety.org/vol9/iss2/art3/

Walton, J. S. (2008) Scanning beyond the horizon: Exploring the ontological and epistemological basis for scenario planning', *Advances in Developing Human Resources*, **10** (2), 147-165.

Wheeler, T. and Von Braun, J. (2013) Climate change impacts on global food security, *Science*, **341** (6145), 508-513.

Wilk, R. (2001) Consuming morality, *Journal of Consumer Culture*, **1** (2), 245-260.

Willer, H., Lernoud, J. and Kemper, L. (2018). The world of organic agriculture 2018: Summary, in *The World of Organic Agriculture. Statistics and Emerging Trends 2018*. Research Institute of Organic Agriculture FiBL and IFOAM-Organics International, pp.22-31.

Worsley, A. and Scott, V. (2000), Consumers' concerns about food and health in Australia and New Zealand, *Asia Pacific Journal of Clinical Nutrition*, **9** (1), 24-32.

Yeoman, I. (2012) *2050 – Tomorrow's Tourist*, London: Routledge.

Yeoman, I. (2013) Tomorrow's tourist and the case study of New Zealand, in J. Leigh, C. Webster and S. Ivanov (eds), *Future Tourism: Political, Social and Economic Challenges*. Abingdon, Oxon: Routledge, pp.161-188.

Young, M.E. and McCoy, A.W. (2016) Millennials and chocolate product ethics: Saying one thing and doing another, *Food Quality and Preference*, **49** (Apr), 42-53.

14 Building Future Scenarios

Introduction

This chapter examines challenges for the future and other emerging ideas in the travel, hospitality and events sectors. The tourism industry is rapidly evolving, and change appears to be the dominant trend. Innovative ideas may serve as a future trend or penetrate an industry to generate a new product or service which becomes the norm and are sought out to maintain a vibrant, growing tourism industry. Fifteen future scenarios are explored with the intent to invoke thought and discussion on how change is embraced. The chapter concludes by suggesting that these are only some of the larger challenges faced by the tourism industry and many more issues need to be considered.

15 Tourism themes for consideration

The following themes highlight challenges for the tourism industry identified by a variety of sources. Presenting these themes allows for further discussion in a structured way and ideal for provoking further discussion.

1. **Maintaining a destination's sustainable tourism development: social, cultural, natural and built resources**

    When forming tourism policies and strategies at any destination, social, cultural, natural and built resources are generally considered when striving to be successful and sustainable in the long term (Hassan, 2000). When developing future scenarios, it is important to strive for sustainability. However, researchers find it challenging to define sustainability relating to tourism development but agree that certain factors should be considered. These include global, physical and environmental impacts; environment concerns; equity; organisation and government policy; education; integrity; authenticity; local control; destination and social impacts (Clarke, 1997). More recently, additional considerations have been added that include a greater focus on conservation, community engagement and community resources in conjunction with other sustainable principles (Pederson, 2016). What is evident in the literature and industry best practices is that all identified factors relating to sustainability should be considered even if they cannot be completely acted upon as discussed in Chapter 10.

An example of an international destination is the Kingdom of Bhutan that asserts sustainable guiding principles (refer to Chapter 13), however, there are other destinations that demonstrate sustainable principles pertaining to social, cultural, natural and built resources. Zanzibar, for example, is an island off the coast of Tanzania that showcases the 'potential of benefit-sharing from sustainable tourism in the Kiwengwa-Pongwe tourism zone' (Makame and Boon, 2008: 93). This case study highlights that when consideration is given to both the made-made and natural environments, positive benefits can flow to all parties concerned and be sustainable in the long-term by achieving collective goals.

2. Concerns for safety and security remain an important issue for the travel and tourism industry

There have always been safety and security concerns in the tourism industry. Dating back to the first travellers in the middle ages (merchants, tax collectors, artists, migrants, students, messengers, military, religious figures, nobles, prostitutes, artists, minstrels, intellectuals, vagrants, beggars), the welfare of life and personal possessions were under real threats due to natural human survival instincts (Ashtor, 2014). Theft among these travellers was common. Fast forward 1000 years and safety and security concerns are still an issue that include terrorism, war, civil unrest, food shortages, increases in populations, exploitation of natural resources, decreasing vegetation areas, reduction in animal habitat, global warming, spread of infectious diseases, government policy, use of technology and online security (Pizam and Mansfield, 2006).

Terrorism for example, is a relative new security concern that has dramatically affected the world and impacts the travel experience of people (Tarlow, 2006) that is also discussed further in Chapter 11. When using air travel for instance, travellers need to consider what to pack in their suitcase and hand luggage. Travellers must pass through a security area that x-rays both people and luggage, test for explosive material, pass through facial recognition cameras and finger-print scanning. Visas, passports and ticket documentation are also required before boarding an aeroplane.

Due to terrorism, travellers will also consider their safety and security at the destination such as the location and style of hotel, the country and geographical location, the stability of the government, local laws, cultural differences and any international warnings. On the official Australian government website (Australian Smart Traveller, 2020) website for example, from perceived safe countries such as Singapore states:

> Singapore overall, exercise normal safety precautions. Exercise common sense and look out for suspicious behaviour, as you would in Australia.

To perceived unsafe countries such as Yemen, the Australian Smart Traveller website (2020) states:

> Yemen overall, do not travel. We advise against all travel here due to the very high risk. If you do travel, you should typically seek professional security advice. Be aware that regular travel insurance policies will be void and that the Australian Government is unlikely to be able to provide consular assistance.

The Australian government online warnings suggest that these concerns can be attributed to terrorism, kidnapping, crime, civil unrest and political tension. Other countries have similar government websites that communicate travel warnings.

3. **Responding to increased interest in the long-term impacts on tourism of climate change and global warming**

> 'Climate change is one of the major issues facing us today and has been described as a threat greater than terrorism' (Hall and Higham, 2005: 1) and 'tourism is obviously related to climate' (Hamilton *et al.*, 2005: 253). It is statements such as these that highlight the importance of the impacts of climate change on the tourism industry. Studies have indicated that most of the research has focused on the impacts at the destination, however Hall and Higham suggest that three areas should be covered including the 'tourism generating region, travel to and from the destination, and the destination itself' (2005: 13). The literature has a common theme – the impact of climate change and tourism is generally focused on weather patterns and geography at the destination. For island locations this generally includes the impacts on the beach, rainfall and water temperature. For instance, coral bleaching occurs globally due to changes in water temperature attributed to climate change (Hoegh-Guldberg *et al.*, 2007). Tourists may choose not to visit a tourist destination due to localised coral bleaching resulting in a dramatic impact on the region. Consequently, the economic and social impacts on tourism due to climate change may not recover.

4. Impact on the travel and tourism industry resulting from a global economic-political perspective

Tourism and general travel are directly affected by the global economic and political perspective in a variety of ways. First, the relationship between the governments of two or more countries can impact the travel experience through visas and entry requirements, as mentioned in Chapter 6. Second, the overall resilience of the economy and therefore the disposable income of travellers has a direct correlation as to which destination is selected. Individuals in countries that have relatively low disposable incomes, such as Papua New Guinea, may have no financial means to travel. Whereas those from countries with high disposable incomes such as Switzerland or Germany may have a greater variety of choice, as residents are less restricted by finances (Buhmann *et al.*, 1988). Travel destinations may also be affected by the overall reputation of facilities offered by the government such as airport access, roads, transport and other facilities.

Third, the political viewpoint of travellers may influence their choice of destination. There are a variety of political systems that governments have adopted including pure communism, socialism, constitutional monarchies, democracy, dictatorships and capitalist-based models. Although all models display both positive and negative aspects, it is either wealth, power or both that are the main driving forces that influence which model a government adopts (Eisenhardt and Zbaracki, 1992). Tourists from a democratic society may not wish to travel to a communist country as viewpoints are too different. Conversely, travellers may decide to visit a different government model to that in their home country to see the impacts on that society. There is evidence that this also occurs from a cultural viewpoint and background including gender and age (Seddighi *et al.*, 2001).

5. Necessity for increased local/regional/national leadership in tourism policy and strategic planning and event management

Strategic planning in tourism policy is essential for the success of a tourist destination or an event. Most researchers and tourism practitioners would concur that one of the most important elements of any tourism strategy is identifying the correct target market/s. In fact, McCabe (2018) suggests that a well thought-through strategy incorporates target markets for the benefit of all stakeholders. The positive flow-on effects should outweigh the negative ones. Most tourism strategic plans and policies are driven by governments offering guidelines, rules and regulations for industry to follow (Hall, 2011). At a local level, it is generally instigated by local councils who assist local tourist operators, attractions, events and festivals. At a regional level, it covers an identified geographic area such as states, provinces or a combination of towns, cities or environmental areas such as mountainous regions. The tourism policies should represent all tourism stakeholders within the identified region and can attract inter-regional visitors. At a national level, it is the responsibility of the government to attract in-bound visitors from other countries. Once the country is chosen, it is up to the regional and local governments to entice visitors to their specific region, a strategy known as the co-opetition concept. Additionally, all policies need to consider the triple bottom line framework (social, environmental and financial impacts) (Heath and Wall, 1991). This will allow for greater diversification of tourism stakeholders with arguably different agendas that help shape policy.

6. Resolving barriers to travel: visas, passports, airline services, fees, and delays

To increase travel capacity and decrease barriers to travel, policy makers in the tourism industry will need to address certain concerns. These include making the process of visa and passport applications easier, improving the facilities and services offered at airports and by airlines, and minimising delays through check-in, security and immigration. The European Union (EU) for instance, makes travelling between member countries comparatively easily as there is no requirement for visas. The EU currency (Euro) is used throughout the EU and 800 language interpreters are available for assistance including travel. Conversely, a foreigner travelling to Russia requires much more official documentation, even where bilateral or multilateral agreements exist between countries. For example, to obtain a visa to enter Russia, travellers require a passport, letter of invitation, additional passport photos, up to a ten-year travel history, detailed itinerary, addresses of individual accommodation locations and evidence of ample finances for the duration of the travel period. Some countries have an electronic visa application process whereas others require duplicate paper documentation which is sent to an individual embassy for approval in person or by return post, which is very time consuming. As criminal identity theft is increasing globally including falsifying passports, there needs to be checks in place to combat criminal activity. Still, the process for obtaining a passport for international travellers is often cumbersome and costly (Salter, 2003).

The travel sector is cognisant of wait times at immigration and some countries have adopted electronic passports fitted with a microchip. This stores all relevant information of the traveller such as facial recognition and travel log times and consequently reduces queuing times (Benco *et al.*, 2007). Citizens re-entering their country of origin often have dedicated immigration gates that are generally quicker for these passengers to pass through than the gates for other nationalities.

Additionally, research has found advantages to reducing queuing times. Nikoue *et al.*, (2015: 1) analysed the queuing time at Sydney International Airport and found:

> Time spent in processing zones at an airport are an important part of the passenger's airport experience. It undercuts the time spent in the rest of the airport, and therefore the revenue that could be generated from shopping and dining. It can also result in passengers missing flights and connections, which has significant operational repercussions. Inadequate staffing levels are often to blame for large congestion at an airport.

Avid air travellers would not disagree that planes are often late in arrivals and departures justified by placing blame on airline traffic control or stating the gate is not available. However, a likely explanation that Mayer and Sanai (2003: 2) suggest is that 'airlines minimalise labour costs at their passengers' expense, although there is some support in favour of airlines trying to maintain greater airline utilisation'. These researchers also suggest that airlines are acutely aware of these issues, but it is tolerable to passengers. Other airline services resulting in barriers to travel include online reservations, ticketing, luggage check-in, seating/comfort, food, beverages, in-flight entertainment, WIFI availability, mileage programs and children policies. All services can attract customer complaints, nonetheless no more so than fees introduced on previously free services such as paying for seat choices, and luggage restrictions (Brueckner *et al.*, 2015).

7. Educating users about optimising the application of new technologies in the travel and hospitality industry

There is evidence that more studies on the factors that influence the adoption of technology in tourism are essential. For instance, Ukpabi and Karjaluoto (2017) analysed 71 peer reviewed journal articles on consumer acceptance of technology in the tourism industry. The findings suggested consumers are embracing tourism apps and mobile technologies, however, the acceptance rate differs depending on their social or cultural groups. Second, a destination needs to create a website and apps that are image conscious, secure, trustworthy and take into consideration privacy concerns of potential customers especially when financial transactions are involved. Third, 'stakeholders must understand the role of website design, social media and mobile technology and e-tourism platforms…TripAdvisor, You-Tube, Lonely Planet, Flickr, Delicious and Digg' that are very important to the tourism and hospitality sectors (Ukpabi and Karjaluoto, 2017: 627). Fourth, Ku and Chen (2015) highlight the importance of interaction; quick navigation and creativity in packaging is essential. Furthermore, satisfaction of the tourist can be improved by the information and content presented and enhanced through audio, video clips and vignettes. By providing accessible, mobile, functional and current content, customers will be more likely to adopt new technologies and have a willingness to be educated in usage, as seen in the success of Facebook that does this well (Barr, 2018). Further discussion can be found in Chapter 3.

8. **Understanding the transformative effect that tourism has on the geopolitics of socio-economic progress**

Development, globalisation and sustainability are the three main impacts on the geopolitics of socio-economic progress (Mowforth and Munt, 2008). Geopolitics relates to the management of the country's resources and the resultant socio-economic progress. The transformation effect can be exponential. Tourism can assist with increasing the wealth of a region economically, nonetheless tourism may also create negative impacts if not fully considered. To highlight the impacts of tourism from a geopolitical viewpoint, Ha Long Bay in Vietnam (Figure 14.1) will be used as a case study.

Figure 14.1: Ha Long Bay, Vietnam. *Source:* P. Strickland, 2020.

Ha Long Bay was listed as a World Heritage Site by UNESCO in 1994. UNESCO has many objectives including the promotion of world peace, preserving culture, heritage and nature. For the residents of Ha Long Bay, for at least the last 20 years, tourism has had a massive positive effect on the economy. Tourism has created jobs, increased wages and encouraged local and foreign investment in the area as explained in Chapter 10.

Additionally, tourism has supported:

> a variety of cultural activities, cultural exchange, recreational activities and restoration of historic buildings…improve residents' living standard and quality of life, foster residents' pride of their local culture, bring good quality of products and services to tourism transportation (Pham, 2012: 36).

However, Pham (2012) also highlighted perceived negative effects identified by the same Vietnamese respondents. For instance,

> tourism was blamed for inflated real estate prices and the increases in the cost of many goods and services; causing traffic congestion, solid waste, air, water, noise and soil pollution; and making natural landscapes and agricultural lands less available (Pham, 2012: 36).

Consistent with Point 5 in this chapter, the residents of Ha Long Bay also have concerns regarding tourism policy. Residents would prefer to be included in development discussions prior to commencement rather than having changes introduced using a top-down approach. Tourism policies must empathise with the environment, to ensure locals will not begrudge tourism and be less welcoming to tourists. Residents would also like to be included in the decision-making process of all tourism strategies so their voices can be heard. This will assist in mitigating any perceived negativities and more likely have the residents accept future changes (Pham, 2012).

9. Effect on travel and tourism from natural/human-induced disasters, health issues, and political disruptions

Broadly, Prideaux (2003) identified three types of disasters that directly affect tourism:

1 Natural disasters (e.g. earthquakes, typhoons, floods, tsunamis, volcano eruptions, avalanches, bushfires and droughts among others);
2 Climate change; and
3 Global epidemics causing associated health issues in humans, fauna or flora.

All three categories have adverse effects: either immediate (natural disasters), progressively worsening situations (global epidemic) and a longer-term perspective (climate change/global warming). All have the same common negative impacts and that is a reduction in tourist numbers affecting tourism operators in the destination communities where the disaster occurred (Stylidis et al., 2014). The recovery timeline depends on which disaster has occurred, how much damage was caused, the cost to rebuild as well as environmental, cultural and social impacts (Tsai *et al.*, 2016).

Maditinos and Vassiliadis (2008) reported the SARS epidemic had a profound negative impact on worldwide tourism. Some important impacts of SARS that Maditinos and Vassiliadis (2008) highlighted were:

- ☐ International tourism arrivals fell;
- ☐ Air travel to areas affected by the advisories decreased dramatically during the epidemic fuelled by the media;
- ☐ Business travel decreased;
- ☐ Tourist arrivals dropped;
- ☐ Major sporting events were relocated to countries that did not have SARS, such as the USA;
- ☐ Chinese restaurants world-wide has a reduction in customers.

From this one health epidemic, SARS had a massive impact on the tourism industries, especially where the source of the outbreak occurred. The legacy and perceived threat to tourists' health can remain for extended periods of time until trust is re-established. Other global examples of tourist health concerns include Mad Cow Disease in the United Kingdom, 1986-2001, Ebola in Western Africa, 2014-2016 and Dengue Fever, that has been detected in over 100 tropical countries (Gubler, 2002) and Covid-19 in 2020. This most recent pandemic has had a devastating effect and may have changed the world forever. At the time of writing, international travel has ceased, domestic travel is limited, people on all continents are self-isolating, meaning staying at home and only leaving their dwellings for essential services such as food, medical and essential work such as health care providers. Covid-19 has spread the virus at such an exponential rate that hundreds of people are dying on a daily basis globally. In response, some countries or regions are in full lock-down meaning people cannot go outside for two-three weeks. If people cannot work from home, many have been laid off, causing stress and anguish about the future. World economies are collapsing but as this is being written during this pandemic, future studies will be published on this topic.

Political and man-made disasters are a definite deterrent to travel. These include, war, civil unrest, terrorism, threat of political persecution based on citizenship or personal beliefs, differing cultures, contradictory traditions, contrasting local laws, mass tourism and under-investment (lack of access, facilities or services) to name a few (Faulkner, 2001). Travel warnings to outbound tourists regarding countries displaying immediate threats such as war are generally given by governments. Whereas other deterrents such as mass tourism or personal beliefs may influence the final choice of destination based on prior knowledge or word-of-mouth (WOM) recommendations.

It is evident that natural or man-made disasters, health concerns or political disruptions have a detrimental effect on tourism numbers. This then has a flow on negative effect on the tourism sector and associated businesses resulting in an economic decline, and perhaps with consequential environmental and social impacts. Further discussion on the short and long term impacts associated with disasters, including natural, health epidemics and terrorism can be found in Chapter 11.

10. Changes in tourism demand resulting from increased travel by emerging nations

'The tourism of the future will include greater demand on the part of citizens of economically emerging nations' (Chambers, 2009: 353). There are many countries and demographic groups that are considered emerging markets. In general terms, emerging marketing in tourism are usually defined by country and in-bound tourism is based on the country under scrutiny. However, emerging markets can also be identified as tourists from other countries. For example, New Zealand's emerging markets are targeted from Asia, in particular, China (Ministry of Business, Innovation and Employment, 2018). However, there are other demographic groups, such as millennials, gay people, solo women, generation Z and super sabbaticals (mentioned in Chapter 3), which are also considered emerging markets.

Chambers (2009) suggested that emerging markets can also be trendsetters. This author eloquently wrote attributes of trendsetters:

> Trendsetters for the tourism of the near future are likely to be well-educated elites who are familiar with travel and comfortable in culturally diverse situations. They will have an understanding of the consequences of global economic development and will better realise that their participation in tourism comes with a cost to communities and environments through which they pass. They will see value in tourism experiences that support principles of environmental sustainability, heritage preservation and cultural diversity, and human equality. This generation of tourists will have greater choice of travel venues and access to considerably more information on which to base their travel plans, and they will be more likely to expect travel experiences that have breadth as well as depth and that provide opportunities for self-improvement as well as leisure and entertainment (Chambers, 2009: 353).

Emerging markets tend to be from a geographic location or a certain demographic. Trendsetters (now known as 'influencers') showcase differing attributes that assist in shaping the desires and hence the travel products and services that should be offered (Mendez, 2019). Both products and services should be considered when shaping tourism policy and marketing strategies.

11. Space tourism and how the industry will operate

The concept of space tourism is not a new phenomenon but certainly an emerging area of research and a new tourism concept that will soon become a reality (Chang, 2015; Crouch, 2005). There are many differing views of how space tourism will take place, however, Collins (1999, also cited in Strickland, 2012: 897) suggested the activity itself will determine the type of space tourist, such as wanting to view the earth, observe the sky, engage in low-gravity sports, observe low-gravity phenomena, swim in low-gravity or artificial-gravity pools, walk in space, spend time in low-gravity gardens, or be immersed in simulated exotic worlds.

It is inevitable that space travel will occur, it is just a matter of when. The 'space race' as it is referred to (Siddiqi, 2000) has been a challenge that many countries have strived to achieve for many decades mainly sponsored by governments. Currently, space tourism is not a commercial reality however space travel is likely to occur in the next few years. This is due to creating new tourism experiences and the potential to be economically feasible. There are many theories as to the first space travellers and most will concur, it will be initially for the wealthy, adventurous and extroverted such as astronauts chosen for the Mars One expedition (Do *et al.*, 2016).

12. Underwater tourism both natural and man-made

As mentioned in Chapter 4, underwater hotels are an emerging trend that will continue to advance. Underwater hotels give a new experience to the guest and are becoming more financially feasible. Water covers most of the Earth therefore it makes sense to explore these frontiers as global populations increase. Hotels however, are not the only facilities and services offered regarding underwater tourism. In the academic literature, underwater tourism fits within the definition of adventure tourism; 'broadly, it means guided commercial tours, where the principal attraction is an outdoor activity that relies on features of the natural terrain, generally requires specialised equipment, and is exciting for the tour clients' (Buckley, 2007: 1428). Buckley also suggests:

> typical activities in adventure tours include climbing, caving, abseiling, sea kayaking, white-water kayaking, rafting, diving, snorkelling, skiing, snow-boarding, surfing, sail boarding, sailing, ballooning, skydiving, parapenting, horse riding, mountain biking, snowmobiling and off-road driving (2007: 1428).

Types of underwater activities listed here include diving and snorkelling although this can be extended to examining marine life, coral reefs and shipwrecks. Other scholars suggest underwater tourism also fits within the realms of marine tourism, scuba diving tourism, sustainable tourism and eco-friendly tourism (Dimmock and Cummins, 2013).

Underwater tourism can be both natural and man-made, which are easy to distinguish with man-made being hotels and shipwrecks whereas everything else is natural. Although underwater tourism is increasing in tourist numbers, there are limitations and boundaries that impede its full potential, including tourists' ability to swim, not having a scuba-diving licence or inadequate finances to stay in expensive hotels as current underwater hotels are targeted towards the wealthy.

13. Food and water security – implications for the hospitality industry

Chapter 13 explores food security in depth, however does not explicitly discuss water security. Fresh water needs to be captured for future consumption. Most countries have invested in dams and large storage reservoirs although this relies on rain falling in the right locations. Continuous droughts have proven that this is not the most efficient method of water storage (Baxter, 1977). Other methods of water harvesting include drilling for bore water, desalination plants and using large water containers.

Gossling *et al.*, (2012) highlighted that water use in the tourism industry needs to be addressed. Over-exploitation of water is rife in hotels and restaurants, generally because customers believe they are paying for the privilege of water use. This is particularly evident in the hotels and the increased time in shower use compared to at home (Lockyer, 2005). To combat this, many hotels are becoming eco-friendly to reduce water consumption. This often entails timers on water taps, restrictive shower heads, reduced water pipe sizes, removal of baths and written reminders for customers to be mindful of water consumption. Restaurants are also reducing water usage and often hotel gardens have been adapted to using grey water (Deng and Burnett, 2002).

14. Will virtual technology reduce the demand for travel?

Virtual reality 'is defined as the use of a computer-generated 3D environment – called a "virtual environment" (VE). Real-time simulation can allow one to navigate and possibly interact with one or more of the user's five senses' (Guttentag, 2010: 637). It has been predicted that virtual technology will result in new products and services affecting the tourism sector. For instance, space travel that is not currently commercially viable could be an opportunity for the tourism industry which could create a simulated environment through virtual reality. This would solve the problem of actual space travel, required training and would reduce costs. Similarly, tourists may pay for a simulated holiday experience rather than travelling to a destination. It is plausible that they will partake in a simulated destination saving time and money. If more products and services in the world of virtual reality are created, it is possible that this will flow into current tourism offerings. This could then have an impact on travel demands. A more in depth discussion of virtual tourism appears in Chapter 3.

15. How will the MICE industry respond to external industry forces?

MICE stands for 'meetings, incentives, conventions, exhibitions' (McCartney, 2008: 293), now often referred to as simply the 'meetings industry'. The MICE industry ebbs and flows depending on external industry forces such as government investment, security, necessity, cost, support from businesses and industry or an individual's disposable income amid others. To remain resilient and be financially viable in the long-term, the MICE industry needs to effectively respond to all challenges even if they have little to no control over changes impacting the industry. Being innovative by creating new ideas and ways of communicating is key to the MICE sector such as hybrid and virtual conferencing that links people globally without the need for travel (Hamm *et al.*, 2018). For example, instant interactive technologies are being introduced during conferences allowing delegates to remotely direct questions to panels, climate control options and catering choices. As this is still a growing sector, innovation and change is key to remaining a financial success.

Summary

The chapter suggests that there are larger challenges faced by the sector and highlights factors that will be drastically affected by change. Building future scenarios in the travel, hospitality and event sectors is essential for long-term viability and financial success. To achieve these objectives, the tourism industry needs to embrace change through new and innovative products and services. The world is rapidly changing, especially with the use of technology and the tourism industry needs to keep up or suffer the negative consequences.

Case study: Demand forecasting and the need for innovative methods

View the clip: https://www.youtube.com/watch?v=HW_bUo1BNDo

Discussion questions

1 Should tourism research guide the tourism industry with future innovation, or should tourism research simply reflect on what is occurring, which is essentially industry driven? Explain your thoughts.

2 How can the tourism industry quickly adapt to changing technologies? Is more financial investment required or should it be consumer led?

Additional resources

- Australian Government Travel Warnings: https://smartraveller.gov.au/Countries/Pages/default.aspx
- European Union: https://europa.eu/
- Importance of the role of academia and the need for policy to drive tourism: https://www.youtube.com/watch?v=pw-XadkWkdo
- Russian visa: https://visalink.com.au/russia-visa
- Sustainability in Bhutan: https://www.youtube.com/watch?v=kMjbru15oCl
- UNESCO: https://en.unesco.org/

References

Ashtor, E. (2014). *Levant Trade in the Middle Ages*, New Jersey: Princeton University Press.

Australian Smart Traveller. (2020) *Yemen*, 22 January, viewed 5/02/2020, smartraveller.gov.au/countries/middle-east/pages/yemen.aspx#safety_and_security

Barr, S. (2018) When did Facebook start? The story behind a company that took over the world, *The Independent*, 23 August, viewed 7/09/2019, https://www.independent.co.uk/life-style/gadgets-and-tech/facebook-when-started-how-mark-zuckerberg-history-harvard-eduardo-saverin-a8505151.html

Baxter, R.M. (1977) Environmental effects of dams and impoundments, *Annual Review of Ecology and Systematics*, **8** (1), 255-283.

Benco, D.S., Mahajan, S., Sheen, B.S. and True, S.L. (2007) *Network Support for Electronic Passports*, U.S. Patent No. 7,221,931.

Brueckner, J.K., Lee, D.N., Picard, P.M. and Singer, E. (2015) Product unbundling in the travel industry: The economics of airline bag fees, *Journal of Economics & Management Strategy*, **24** (3), 457-484.

Buckley, R. (2007) Adventure tourism products: Price, duration, size, skill, remoteness, *Tourism Management*, **28** (6), 1428-1433.

Buhmann, B., Rainwater, L., Schmaus, G. and Smeeding, T.M. (1988) Equivalence scales, well-being, inequality, and poverty: sensitivity estimates across ten countries using the Luxembourg Income Study (LIS) database, *Review of Income and Wealth*, **34** (2), 115-142.

Chambers, E. (2009) From authenticity to significance: Tourism on the frontier of culture and place, *Futures*, **41** (6), 353-359.

Chang, Y.W. (2015) The first decade of commercial space tourism, *Acta Astronautica*, **108** (Mar-Apr), 79-91.

Clarke, J. (1997) 'A Framework of Approaches to Sustainable Tourism', *Journal of Sustainable Tourism*, **5** (3), 224-233.

Collins, P. (1999) Space activities, space tourism and economic growth, *2nd International Symposium on Space Tourism*, 14 May, Bremen, Germany.

Crouch, D. (2005) Flirting with space-tourism geographies as sensuous/expressive practice, in C. Cartier and A.A. Lew (eds), *Seductions of Place: Geographical perspectives on globalization and touristed landscapes*, London: Routledge, pp.23-35.

Deng, S. M. and Burnett, J. (2002) Water use in hotels in Hong Kong, *International Journal of Hospitality Management*, **21** (1), 57-66.

Dimmock, K. and Cummins, T. (2013) History of scuba diving tourism, in G. Musa and K. Dimmock (eds), *Scuba Diving Tourism*, London: Routledge, pp.32-46.

Do, S., Owens, A., Ho, K., Schreiner, S. and De Weck, O. (2016) An independent assessment of the technical feasibility of the Mars One mission plan–Updated analysis, *Acta Astronautica*, **120** (Mar-Apr), 192-228.

Eisenhardt, K.M. and Zbaracki, M.J. (1992) Strategic decision making, *Strategic Management Journal*, **13** (S2), 17-37.

Faulkner, B. (2001) Towards a framework for tourism disaster management, *Tourism Management*, **22** (2), 135–147.

Gossling, S., Peeters, P., Hall, C.M., Ceron, J.P., Dubois, G. and Scott, D. (2012) Tourism and water use: Supply, demand, and security. An international review, *Tourism Management*, **33** (1), 1-15.

Gubler, D.J. (2002) Epidemic dengue/dengue hemorrhagic fever as a public health, social and economic problem in the 21st century, *Trends in Microbiology*, **10** (2), 100-103.

Guttentag, D. (2010) Virtual reality: Applications and implications for tourism, *Tourism Management*, **31** (5), 637-651.

Hall, C. M. (2011) A typology of governance and its implications for tourism policy analysis, *Journal of Sustainable Tourism*, **19** (4-5), 437-457.

Hall, C.M. and Higham, J. (2005) Preface, in C.M. Hall & J. Higham (eds), *Tourism, Recreation and Climate Change*, London: Channel View Publications, pp.1-3.

Hamilton, J.M., Maddison, D.J. and Tol, R.S.L. (2005) Climate change and international tourism: A simulation study, *Global Environmental Change* **15** (3), 253–266.

Hamm, S., Frew, E. and Lade, C. (2018) Hybrid and virtual conferencing modes versus traditional face-to-face conference delivery: a conference industry perspective, *Event Management*, **22** (5), 717-733.

Hassan, S.S. (2000) Determinants of market competitiveness in an environmentally sustainable tourism industry, *Journal of Travel Research*, **38** (3), 239-245.

Heath, E. and Wall, G. (1991) *Marketing Tourism Destinations: A strategic planning approach*, New York: John Wiley and Sons.

Hoegh-Guldberg, O., Mumby, P J., Hooten, A J., Steneck, R.S., Greenfield, P., Gomez, E. and Knowlton, N. (2007) Coral reefs under rapid climate change and ocean acidification, *Science*, **318** (5857), 1737-1742.

Ku, E.C. and Chen, C.D. (2015) Cultivating travellers' revisit intention to e-tourism service: the moderating effect of website interactivity, *Behaviour and Information Technology*, **34** (5), 465-478.

Lockyer, T. (2005) Understanding the dynamics of the hotel accommodation purchase decision, *International Journal of Contemporary Hospitality Management*, **17** (6), 481-492.

Maditinos, Z. and Vassiliadis, C. (2008) Crises and disasters in tourism industry: Happen locally – affect globally, in *MIBES E-Book*, pp.67-76, viewed 6/07/2020, http://mibes.teithessaly.gr/ebook/ebooks/maditinos_vasiliadis%2067-76.pdf

Makame, M.K. and Boon, E K. (2008) Sustainable tourism and benefit-sharing in Zanzibar: the case of Kiwengwa-Pongwe Forest Reserve, *Journal of Human Ecology*, **24** (2), 93-109.

May, K. T. (2018) Why we should all consider taking a midlife gap year, *TED Ideas*, 2 November, viewed 10/06/2019, https://ideas.ted.com/why-we-should-all-consider-taking-a-midlife-gap-year/

Mayer, C. and Sinai, T. (2003) Why do airlines systematically schedule their flights to arrive late? Pennsylvania: The Wharton School, University of Pennsylvania.

McCabe, S. (2018) Stakeholder engagement in tourism, in L. Moutinho and A. Vargas-Sanchez (eds), *Strategic Management in Tourism*, Oxfordshire: CABI Tourism Texts, pp.279-294.

McCartney, G. (2008) The CAT (casino tourism) and the MICE (meetings, incentives, conventions, exhibitions): Key development considerations for the convention and exhibition industry in Macao, *Journal of Convention & Event Tourism* **9** (4), 293-308.

Mendez, L. (2019) Inside the complicated world of the travel influencer, *CNN Travel*, 2 September, viewed 2/09/2019, https://edition.cnn.com/travel/article/travel-influencers/index.html

Ministry of Business, Innovation and Employment. (2018) New Zealand Tourism Forecasts 2018-2024, viewed 5/02/2020, https://www.mbie.govt.nz/assets/5c05b7bfce/nz-tourism-forecasts-2018-2024-report.pdf

Mowforth, M. and Munt, I. (2008) *Tourism and Sustainability: Development, Globalisation and New Tourism in the Third World*, London: Routledge.

Nikoue, H., Marzuoli, A., Clarke, J. P., Feron, E. and Peters, J. (2015) Passenger flow predictions at Sydney international airport: a data-driven queuing approach, 20 August, viewed 5/02/2020, https://arxiv.org/abs/1508.04839

Pederson, A. (2016) Frameworks for tourism as a development strategy, in S.F. Cool and K. Bosak (eds), *Reframing Sustainable Tourism*, Dordrecht, Netherlands: Springer, pp.47-63.

Pham, L. (2012) Tourism impacts and support for tourism development in Ha Long Bay, Vietnam: An examination of residents' perceptions, *Asia Social Science*, **8** (8), 28-39.

Pizam, A. and Mansfeld, Y. (2006) Toward a theory of tourism security', in Y. Mansfeld & A. Pizam (eds), *Tourism, Security and Safety: From Theory to Practice*, London: Routledge, pp.1-28.

Prideaux, B. (2003) The need to use disaster planning frameworks to respond to major tourism disasters: Analysis of Australia's response to tourism disasters in 2001, *Journal of Travel & Tourism Marketing*, **15** (4), 281-298.

Salter, M.B. (2003) *Rights of Passage: The passport in international relations*. London: Lynne Rienner Publishers.

Seddighi, H.R., Nuttall, M.W. and Theocharous, A.L. (2001) Does cultural background of tourists influence the destination choice? An empirical study with special reference to political instability, *Tourism Management*, **22** (2), 181-191.

Siddiqi, A.A. (2000) *Challenge to Apollo: the Soviet Union and the space race, 1945-1974*. Washington, USA: NASA.

Strickland, P. (2012) Do space hotels differ from hotels on earth? The mystery is solved, *Journal of Hospitality Marketing & Management*, **21** (8), 897-908.

Stylidis, D., Biran, A., Sit, J. and Szivas, E.M. (2014) Residents' support for tourism development: The role of residents' place image and perceived tourism impacts, *Tourism Management*, **45** (Dec), 260-274.

Tarlow, P.E. (2006) A social theory of terrorism and tourism, in Y. Mansfeld & A. Pizam, (eds.), *Tourism, Security and Safety: From Theory to Practice*, London: Routledge, pp.29-32.

Tsai, C.H., Wu, T.C., Wall, G. and Linliu, S.C. (2016) Perceptions of tourism impacts and community resilience to natural disasters, *Tourism Geographies,* **18** (2), 152-173.

Ukpabi, D.C. and Karjaluoto, H. (2017) Consumers' acceptance of information and communications technology in tourism: A review, *Telematics and Informatics,* **34** (5), 618-644.

15 Summary

In this book we have examined the current and future capabilities of the tourism, hospitality and events industry by exploring the opportunities available to shape the future through rebuilding, disrupting and developing greater resilience in the tourism industry. We set out in writing the book in times when there was economic prosperity and stability, however this changed all too quickly with the advent of Covid-19.

Over the past months the developments across the world have resulted in unprecedented change and disruption, particularly to the travel, hospitality and events sectors. Yet we firmly believe the tourism industry will survive, however we acknowledge that it will take some time to rebuild traveller and consumer trust and there will be an indelible impact on business operators and the industry overall. As we have emphasised throughout the book, it becomes more important than ever to evaluate the future realising that change is inevitable and that there will be peaks and troughs within industries that have to be managed over time.

Three features of futures studies set the tone of this book. First, a systems view of the industry was adopted allowing for a holistic understanding of the scale of industry, and the important inter-relationships existing between stakeholders likely to shape future scenarios. Second, the potential and probable future trends based on an analysis of the socio-cultural technological, economic, environmental, political and international dimensions were examined. Finally, a medium to long-term view of the future potential and opportunities available to the tourism industry was considered.

As an open system, the tourism industry is subject to constant change and an array of subsequent impacts. While past drivers of change have opened the world up to increased global travel, current and new drivers will shape future travel demand as well as industry operations and supply. Societal changes and shifting tourist demand will result in more personalised and sustainable experiences being sought with the role of technology in facilitating these experiences becoming more essential.

Increased growth in travel and the demand for more personalised experiences has resulted in some destinations not being able to cope. This intensifies the need for effective destination management to ensure that resources are used efficiently, and that biodiversity conservation is promoted, and ultimately, the negative impacts of tourism will be consciously reduced. We as tourists are seeking new cultures and landscapes that often involve long haul travel. Subsequently, cleaner and more efficient transport alternatives will be required to satisfy both this increase in demand for new experiences along with the desire of tourists to be more sustainability conscious.

New and emerging technologies will continue to impact the travel, hospitality, and event sectors greatly. Travellers will expect digital connectivity, and smart technology makes it possible for tourists to participate in and enhance their own visitor experience. Technological change challenges traditional employment models, particularly within the hospitality sector, especially through the adoption of robots to perform manual tasks that were traditionally performed by human staff. Substituting human labour with electronic robots, however, raises questions regarding their ability to deliver quality, trustworthy and exceptional customer service. Future events are likely to be impacted by contemporary issues of sustainability, event inclusivity and event technology. Event organisers will be required to balance the sustainability element with the increasing demand for face-to-face interaction. While sustainability in the past was viewed as a niche element of tourism, it now must become the 'new norm' for all sectors operating within the international tourism industry (UNWTO, 2020a).

The changing nature and scale of crises will require an increased understanding of industry relationships and the way such change and its impacts is managed by the various sectors. The COVID-19 crisis outbreak reinforces the vulnerability of the international tourism industry operating as an open system and highlights the impact of change on future industry development. One hundred percent of destinations worldwide have imposed some form of travel restrictions, with the first 2020 quarter data indicating a 22% decrease in international tourist arrivals, representing a loss of 67 million international arrivals, and an estimated USD $80 billon lost in exports during this period when compared to the same period of the previous year (UNWTO, 2020b). Reduced travel restrictions placed on business operations along with decreased personal interactions has resulted in the tourism industry been brought to a standstill (OECD, 2020). Many governments have introduced economy wide measures, including tourism specific stimulus packages, with the aim of minimising industry impacts and aiding the recovery process of rebuilding destinations (OECD, 2020).

Both the OECD (2020) and UNWTO (2020b) have released impact assessments based on industry expert evaluations of the potential effects of COVID-19 on international tourism using three scenarios. These scenarios reflect three possible patterns of monthly changes in tourist arrivals for the remainder of 2020, based on when travel restrictions are lifted: Scenario 1 sets the date at early July 2020; Scenario 2 sets it at early September 2020; and Scenario 3 set it at early December 2020. A reduction in tourist arrivals of between 58-60% was identified for scenario 1, 70-75% for scenario 2 and 78-80% for scenario 3 (OCED, 2020; UNWTO, 2020b). It is expected that the European Union regions will experience an earlier recovery than other global regions although a later and slower timeframe than initially anticipated is expected overall (OECD, 2020). The rate of recovery will ultimately be dependent upon when travel restrictions are lifted globally, although a panel of experts are inclined to suggest a start of the recovery process of international tourism demand won't occur until in 2021 (UNWTO, 2020b). Governments are preparing comprehensive tourism recovery plans that will need to incorporate new safety and health protocols. International traveller confidence will take time to be restored and the economic flow on effects of the pandemic will restrict many people initially from international travel. Meanwhile domestic tourism, which accounts for approximately 75% of tourism in OECD countries, offers the primary means of driving destination recovery (OECD, 2020). This does not take into consideration a second wave of economic and mass people movement shutdowns due to an increase in coronavirus cases, that have been witnessed in countries such as China, Singapore and Australia.

Although the Covid-19 crisis has presented the international tourism industry with complex and challenging times of uncertainty, tourism can be used as a platform for pandemic recovery, it provides a means of bringing people together and can promote solidarity and trust at a time when it is most needed (Guterres, 2020). Opportunity exists for destinations to diversify their markets and encourage innovation and investment. 'To build a stronger, more sustainable and resilient tourism economy, the crisis is an opportunity to rethink tourism for the future' (OECD, 2020).

Several chapters in this book have highlighted particular events that have had major impacts on the industry, often for sustained periods of time. Events such as wars, pandemics, financial crises, new technology climate change and changing trends and attitudes can all have dramatic impacts on industries. We have aimed to present several approaches to preparing for such significant industry events including considerations for sustainability, crisis management, systems thinking, design thinking and scenario planning. Taking the latter approach, we can identify six considerations to help in preparing for the future.

1. Plan for uncertain events that can impact the business;
2. Consider the implications on staff and customers;
3. Ensure that current procedures and practices are capable of handling extenuating circumstances likely to occur in a crisis;
4. Allocate resources (time, effort and funds) to the preparation and testing of responses to unforeseen events;
5. Conduct regular training of staff and role plays for unforeseen events; and
6. Build resilience through sharing and connections with networks and external organisations.

At the end of the day planning for the future is essential whether it be for good times or bad. This will have to be the new normal of the industry as the future will always be unpredictable, however it may be manageable if planning for the unknown is always considered.

While tourism is generally one of the first industries to suffer impacts of economic adversity, it is an industry which has also proven its resilience in recovery from other past global crises events, including wars, pandemics, SARS and the Global Financial Crisis (Tourwriter, 2020). The dynamic nature of the tourism industry requires constant analysis of probable future trends in order to best plan and manage the associated impacts of change as 'the measures put in place today will shape tourism of tomorrow' (OECD, 2020).

References

Guterres, A. (2020) Tourism can be a platform for overcoming the pandemic. By bringing people together, tourism can promote solidarity and trust, *UNWTO*, 9 June, viewed 15/06/2020, https://www.unwto.org/news/tourism-can-promote-solidarity-un-secretary-general-antonio-guterres

OECD. (2020) *Tourism Policy Responses to the Coronavirus (COVID-19)*, 2 June, viewed 14/06/2020, https://www.oecd.org/coronavirus/policy-responses/tourism-policy-responses-to-the-coronavirus-covid-19-6466aa20/

Tourwriter. (2020) *The travel industry is resilient: How to survive economic adversity*, n.d. viewed 15/06/2020, https://www.tourwriter.com/travel-software-blog/covid-19-pt2/

UNWTO (2020a) *International tourist numbers could fall 60-80% in 2020, UNWTO reports*, 7 May, viewed 20/05/2020 https://www.unwto.org/news/covid-19-international-tourist-numbers-could-fall-60-80-in-2020

UNWTO (2020b) *UNWTO Tourism Data Dashboard*, n.d., viewed 1/06/2020, https://www.unwto.org/unwto-tourism-dashboard

Index

ageing population
  tourism opportunities 9
  wellness tourism 127
agricultural resources 190–191
agriculture, impact on environment 190
air travel 194–195
  and climate change 12
  impact of Covid-19 11
animal ethics 15
augmented reality 25
  dining experiences 26–27

barriers to travel 209
beacon technology 24
beauty spa hotels 122
Bhutan, limits on tourism 192
Brundtland Commission 136
bucket list 58–61
business events 76–78

Case studies
  2028 Los Angeles Olympics 81–83
  Business Council of Australia scenario summaries 2025 182–183
  Demand forecasting and the need for innovative methods 218
  Food security 200
  Future industry training and development 98

  iSimangaliso Wetland Park, South Africa 143–145
  Personalising the travel experience 32–34
  The Game of Thrones tourist phenomenon 111–113
  The impact of terrorism on France as a popular tourist destination 160–164
  Village Roadshow Theme Parks 66–68
  Wellness spa tourism in Thailand 128–130
causal mapping 176
climate change 207
co-creation 23
collaborative consumption 14
Complementary and Alternative Medicine 119
consumer activism 15
Covid-19 155–156, 213
  impact on tourism spending 1
creative tourism 30
crises 149–170
  categories 149
  definition 149
  impact on international tourism industry 151–157
crisis management
  definition 150
  future approaches 158–159
  models and strategies 156–157

Crocodile Dundee marketing campaign 110
cruise ship industry 195–196
  crowth 41
  impact of Covid-19 43
cultural tourism 103

dark tourism 58
Delphi Method 180
design thinking 181
destination images and film-induced images 105
destinations, provision of photo opportunities 61
digital detox 27
digital marketing, personalised 32
digital natives 13
Digital/Open Badges 92
disasters 212
disease outbreaks 155–156
drivers of change 7–20

emerging markets 214
escape rooms 58
events 73–86
  active engagement 77
  Covid-19 80
  inclusivity 76
  networking 77
  reasons for gatherings 73
  safety and security 80
  sustainability 75–76
  venues 77

facial recognition to replace tickets 57
film cruise 109
film tourism 103–116
  benefits 105–106
  challenges 107–108
  virtual reality 108
flygskam (flight shaming) 12

FOMO (Fear of Missing Out) 14, 74
food security 196–198, 216
  and tourism 198
  case study 200
forest therapy 126
France, impact of terrorism 160–164
fuels used in tourism 189–190
future scenarios 205–222

gap year 28
  volunteer activities 29
gathering, defined 73
generational cliques 13–14
Generation Z 13
geopolitics 211
globalisation 11
global warming 207
growth of tourism, factors influencing 8

health tourism 117–134
heuristics 172
hospitality 39–54
  customer service 43–47
  key functions of the sector 39
  use of robots in Asia 44–50
human nature and travel 10–11
Hurricane Katrina 154

inclusivity, events 76
Indian Ocean Tsunami 154
Instagram
  and overtourism 61–65
International Centre of Excellence in Tourism and Hospitality Education 89

JOMO (Joy of Missing Out) 14

leisure
  relationship with work 8–10
  retirees 9
lifestyle resorts 122

local environment, impact of tourism 191
locations
  and overtourism 107
  benefits to property owners 106
  organised tours 109
locations, filming 103–105

marketing campaign
  Son of Crocodile Dundee 110
Maslow's Hierarchy of Needs 40
Massive Open Online Courses 91–92
medical tourism, defined 118
Meetup online platform 74
meme tourism 59
MICE industry 217
MICE sector 76
mixed reality 25
morality in tourism 191–192

natural disasters, impacts 153–155
natural environments, protection 139
natural resources 188–191
  the tourism 188–192
networking, events 77

Open Badges 93–97
  in higher education 94–95
overtourism 61–65, 195
  responses to 64–65
  top 10 tourist cities 62

pandemics 155–156
personalisation
  airlines and hotels 32–34
  geo-targeting 33
plausible futures 187–190
population growth 194, 194–195
problem solving 171–186
  heuristics 172
  non-linear solutions 173
  traditional approaches 171–174

queues,
  express 57
  virtual 57

responsible tourism 135–148
  systems approach 141–142
  whose responsibility? 140–141
retirement age 9
robots in hospitality
  advantages and disadvantages 48–49
  main benefits 45
runaway films 104

safety and security 206
scenario planning 178–181
scenarios 205–222
  developing 179
selfies
  accidental deaths 60
  and social comparison theory 60
  on social media 59
September 11 attack, long-term impact 153
smart boredom 27
smart business 23
smart destinations 22
smart tourism 22–24
  defined 22, 23
social media 23
  posting selfies 59
solo travellers 30–31
space hotel employees
  skills required 41
space tourism 215
space travel 40
spa tourism, defined 118
spiritual retreats 123
spiritual tourism 126
stocks and flows tool 177–178
strategic planning 208

super sabbatical
  creative tourism 30
  volunteer activities 29
super sabbaticals 28–30
sustainability
  events 75–76
  social, economic & environmental 138–139
sustainable development 135–148
Sustainable Development Goals and tourism 137–140
sustainable tourism development 205
systems thinking 175–178
  causal mapping 176–177
  defined 175
  stocks and flows 177–178

tagskryt (train bragging) 12
technologies, application of new 210
terrorism 151–153
  defined 152
  impact on French tourist industry 160–164
terror parks 58
THE education
  and ICT 90
  development 88
  international students 89
theme parks 55–58
THE teaching and training 87–102
time-poor and time-rich tourists 9
tourism
  defined 22
  industry value, 2019 2
  spending, impact of Covid-19 1
tourism demand, emerging markets 214

tourism industry, definition 2
tourist arrivals 2019 21
  expected growth 135
tourist attractions
  must-see 61
tourist origins 7
Tuning Project 89

underwater hotels 41, 215
UNESCO Heritage sites 63, 211
User Experience Design (UX) 181

venues 77
virtual hotel tours for booking 25
virtual queuing 57
virtual reality 25
virtual technology 216
virtual tourism 25–27
virtual travel experiences 25
visitor attractions 55–72
  types 55

Walt Disney World Resort 56
war and tourism 193–194
wellness tourism 117–134
  ageing population 127
  defined 117, 118
  development 120–121
  impact of Covid-19 127–128
  providers 122–124
  trends 124–127
wellness tourist zones 126
White Island volcanic eruption 154
wicked problems 174
work and leisure, relationship 8–10